PETER L. SMITH

Discovering Canals in Britain

SHIRE PUBLICATIONS LTD

ACKNOWLEDGEMENTS
The author acknowledged the assistance he received from Jeremy G. Simpson, Alan West, the Barge and Canal Development Association and from officials of the waterway restoration bodies who supplied useful information.

The publishers acknowledge the valuable assistance of Tony Conder, Curator of The National Waterways Museum, in the preparation of the 1997 reprint.

Photographs are acknowledged as follows: *Waterways News*, page 24; W. G. Brewer, page 62; Barge and Canal Development Association's Museum Loan Service, pages 40 and 42 (top); Royal Armouries Museum, page 68; Jeremy G. Simpson, pages 93 and 101; Josiah Wedgwood and Sons Ltd, page 42 (bottom); Cadbury Schweppes Ltd, page 37 (top); *Daily Mail*, page 37 (bottom); Christopher S. Walker, page 78; Cadbury Lamb, cover, pages 10, 13, 23, 26, 27, 33, 55, 56, 60, 65, 68, 71, 74, 83, 92, 97, 100, 103, 104, 105, 106 and 108. Other photographs are from the collection of the late Peter L. Smith. The maps and line drawings are by John W. Holroyd.

The cover photograph is of Marsworth Dry Dock on the Grand Union Canal near Tring in Hertfordshire.

Published in 1997 by Shire Publications Ltd, Cromwell House, Church Street, Princes Risborough, Buckinghamshire HP27 9AA, UK.
 First published 1981; second edition 1984, reprinted 1986; third edition 1989; fourth edition 1993; reprinted with amendments 1997. Number 257 in the Discovering series. ISBN 0 7478 0204 1.

Printed in Great Britain by CIT Printing Services, Press Buildings, Merlins Bridge, Haverfordwest, Pembrokeshire SA61 1XF.

Contents

Introduction

Discovering the canals of Britain is an activity that many thousands of people participate in every year. Their introduction to the inland waterways may be a short trip on a passenger boat that brings about a realisation that there is so much to see, or it may be an invitation to join some friends on a boating holiday with the prospects of exploring new areas. It could well be because of a desire to do something different on a family holiday, when every member of the family can participate fully. Or indeed it could be because of a pleasant towpath walk in unfamiliar surrounds.

The inland waterways have many changing facets, offering pleasures that few have felt more keenly. There are many hundreds of miles of towpaths waiting to be explored and several hundred miles of waterways under restoration which require the assistance of able-bodied people. Those who enjoy boating can find pleasure on the thousands of miles of quiet idyllic waterways that are available for cruising. This can be done by canoe, as a passenger on a hotel boat, or with skipper and crew of one's own choosing on a hire boat, or on your own craft.

Do explore the canals and, if it is for the first time on a hire cruiser, do study maps beforehand to ensure that you plan your route. Remember that at a relaxing speed of 4 mph (6.4 km/h) you have time to look about you as you cruise along, with ample time to eat your meals comfortably, do a little exploring and finish with a glass in the local in the evening. It is a temptation to try to cover too great a distance. Generally it will take about ten minutes for a narrow lock and so allowing for about eight hours cruising a day you are able to travel some 20 miles, passing through ten locks. The hire rates are much lower out of the popular cruising months. Worth considering are April and May, with the beauty of Spring, and late September and October, which can offer so much with the rich reds and browns of autumn. Do remember some warm clothes for the evening and a radio to listen to while relaxing.

There is plenty for the children to do, for they can help, even in the galley. Those without a family group could try a hotel boat, where they will find some likeminded folk with whom to spend an enjoyable week. Those intending to buy a boat would be wise to have at least a weekend cruise first.

I have tried to give as much information as is possible in a book of this size about the early days of waterways, their development and present use, along with details of some of the most popular sights to see and a brief description of the popular waterways. I hope that I have been able to generate sufficient interest to stimulate the reader to discover more about the canals of Britain.

1. River navigations

In Britain most of the rivers have been used as water highways since early times, although at first little use was made of them because the people lived mainly in self-supporting communities and trade was generally conducted at local markets and fairs. The Romans, during their occupation, were the first people to use inland water transport as we know it. Previously only rudimentary types of craft, such as dug-out canoes made from tree trunks, had been used for short journeys, but the Romans used rafts and boats for longer journeys. They also constructed the first man-made canal in Britain, the Fossdyke, which was used for drainage as well as navigation. They also made the Car Dyke from Lincoln to Peterborough and the Itchen Dyke connecting Winchester with the south coast. The Fossdyke provided a water link between Lincoln and the river Trent at Torksey, and the Romans used it to transport grain grown in the Lincoln area to supply their garrison at York. The loaded boats had to journey along the Fossdyke and the rivers Trent and Ouse, and this must have been a hazardous journey, for on tidal waters the boats had to contend with wind, shoals and strong currents.

After the fall of the Roman Empire and the withdrawal of the Romans from Britain, many centuries were to elapse before there was any further practical improvement of the rivers for the benefit of inland water transport. But then came a demand for stone to build castles, fortified mansions and churches, which resulted in an increased use of certain rivers. This brought about a gradual awakening to the possibilities of water transport and some practical steps forward were made. One of the first improvements was in 1120, during the reign of Henry I, when the Fossdyke was scoured and once more made navigable after it had become silted up. This was followed by improvements to the river Lee in the mid 1420s, the Essex Stour in 1505 and the river Welland in 1571.

The tidal estuaries were used increasingly by sailing barges. On occasions these made journeys inland along rivers that rarely saw the passage of a vessel. The vessels travelled mostly with the tides, relying on the extra water to lift them over fords and shallows, and the further inland the journey went the more arduous and slow it became. On the rivers furthest from the coast journeys generally were only practicable under favourable conditions and depended upon the natural suitability of the river. Craft were hindered in turn by shallows and rapids, while floods in winter and drought in summer also delayed them. Progress depended upon sufficient water to float the craft and enough men to haul them against the fast flowing currents and flood water.

Throughout the middle ages, commerce and the resulting river trade gradually increased to transport building materials, foodstuffs and agricultural products to the settlements that grew into towns and cities along the riverbanks. Still little or nothing was done to the majority of rivers to aid navigation. Usually the lower tidal sections were freely navigable, but above this the owners of the land alongside mainly had control over the river, and they imposed certain restrictions or charges. In many places flour mills were built on the banks, and weirs were then constructed across the rivers to dam the flow of water to provide sufficient pressure to power waterwheels for the mills. These weirs also held the water back so that it provided a suitable depth for navigation for deep-draughted vessels, but the weirs stopped any progress for vessels until the practice of using weirs that incorporated a movable section called a flash lock was established. The flash locks were gateways in the weirs, made up of baulks of timber that slotted into position on top of each other. To let a vessel pass, the baulks of timber were lifted out, and for a vessel going downstream this was usually satisfactory, for it went with the flow of water, but vessels going upstream required teams of men hauling on ropes from the riverbank. Most of the weirs were constructed and owned by the mill owners, and as they were dependent upon a head of water to power their mills they were reluctant to lose it to let a vessel pass, especially in times of drought. Their actions delayed the passage of craft and this resulted in conflict between the millers and the owners of the vessels using the rivers. This state of affairs lasted a long time and was detrimental to inland water transport, as journeys by river continued to be slow and difficult, with many delays, often of several weeks and even longer in periods of drought.

During the eighth century locks similar to those in use today were used for the first time. These, fitted with guillotine gates, were used on the Grand Canal in China, while in the western world the first locks with guillotine gates were built in the 1370s on the river Lek in Holland. In the late fifteenth century Leonardo da Vinci is reputed to have designed the first lock with the now familiar mitred swinging gates. The first lock in Britain using these swinging gates was built in 1566 on a short length of canal that made the river Exe navigable to Exeter. The lock, which provided a practical and safe method of overcoming changes in water level, set the pattern that was eventually to revolutionise inland water transport. From then on throughout Britain locks fitted with mitred gates that close to form a slight V against the pressure and higher level of water were constructed. At first locks were built at the side of the weirs, then at the end of a short cut. These cuts, which were short man-made lengths of canal, were

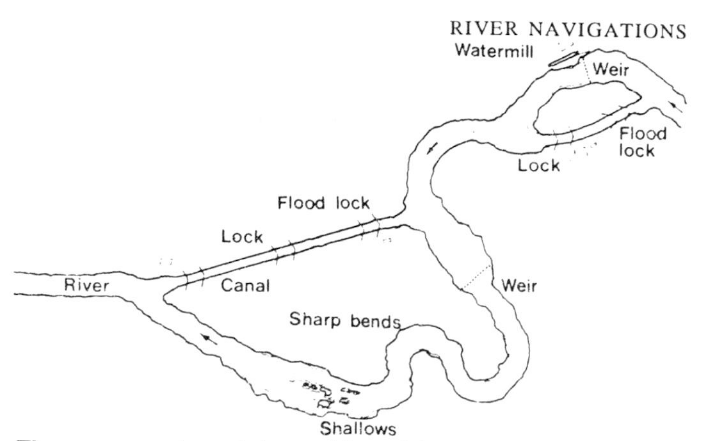

The water supply and the position of the locks on a typical early river navigation.

built to bypass the existing weirs and new ones that were made to improve the navigable channel. The locks, with a pair of gates at either end just far enough apart to accommodate the normal sized vessel in general use, were fitted with sluices to adjust the water level by the difference between the levels above and below the nearby weir. They became commonly known as pound locks, while the lock-free section of water between two separate locks became known as a pound.

The transport requirements of merchants and traders dictated the amount of development in the years to follow. This was usually undertaken by groups of businessmen who considered the profits to be derived from tolls and formed companies to promote works to provide an improved transport system. During the seventeenth century this resulted in the first major expansion and many improvements. Certain rivers, like the river Severn, became exceptionally busy waterways with much financial gain for those involved. Other rivers that were greatly improved at that time included the river Thames to Oxford, the Hampshire Avon, the Great Ouse and the river Wey. Throughout the next century the making of river navigations continued, using a combination of natural rivers and artificial cuts. Initially as much of the natural river as possible was incorporated to save money on making new canal sections, but later some navigations had canal sections many miles in length to bypass shallows and winding stretches of river. These were made possible by improved engineering techniques and these new skills also provided artificial water supplies so that canals were made independent of rivers.

2. The canal age

Before the late eighteenth century inland transport away from the navigable rivers was slow, unreliable and expensive, because of the lack of made-up roads. Horse-drawn carts were able to move over the dry and dusty cart tracks during the summer months, but in winter the thick mud made it impossible. Packhorses could travel almost all the year round, but with their limited carrying capacity they were not an ideal method of transport, especially for fragile, heavy or bulky materials. No guarantee of delivery could be made for either method as they were both subject to the actions of freebooters. Despite these problems commerce and industry were slowly developing, and the time was ripe for a suitable alternative method of transport, but the building of canals was slow to start.

The first modern canal in England was the St Helens Canal, which received an authorising Act for the building of a navigation in 1755. It had been decided to construct a waterway, based on the Sankey Brook, because a number of turnpike roads, levying tolls where previously there had been none, meant that coal from the collieries that was intended for Liverpool and the salt works near the river Weaver would be more expensive and demand would drop. As was common practice at that time, the plan was to canalise the local natural watercourse, the Sankey Brook, which ran from St Helens to the northern shores of the river Mersey. The engineer, Henry Berry, decided that such a plan was impracticable but suggested to one of the proprietors that a canal was possible, receiving a water supply from the Brook. They agreed to build a canal but not to divulge details to the other proprietors, as it was felt that they would be against such a revolutionary scheme. Work commenced on the artificial waterway in September 1755, and two years later the waterway, 10 miles long (16 km) with ten locks, was practically complete.

However, the third Duke of Bridgewater, who had collieries at Worsley near Manchester, is usually given the credit for the first modern English canal. He used packhorses to transport coal from Worsley to Manchester and as a result it was a scarce and expensive commodity in the city. John Gilbert, his agent, pressed the Duke to build a canal and finally convinced his employer that water transport would be a satisfactory method of overcoming their problems. As a result an application was made to Parliament to construct a canal from Worsley to the Manchester outskirts at Salford. In 1759 parliamentary agreement was secured for the Bridgewater Canal on the understanding that coal would be offered for sale in Manchester at 4 d (1.5 p) or less per hundred-

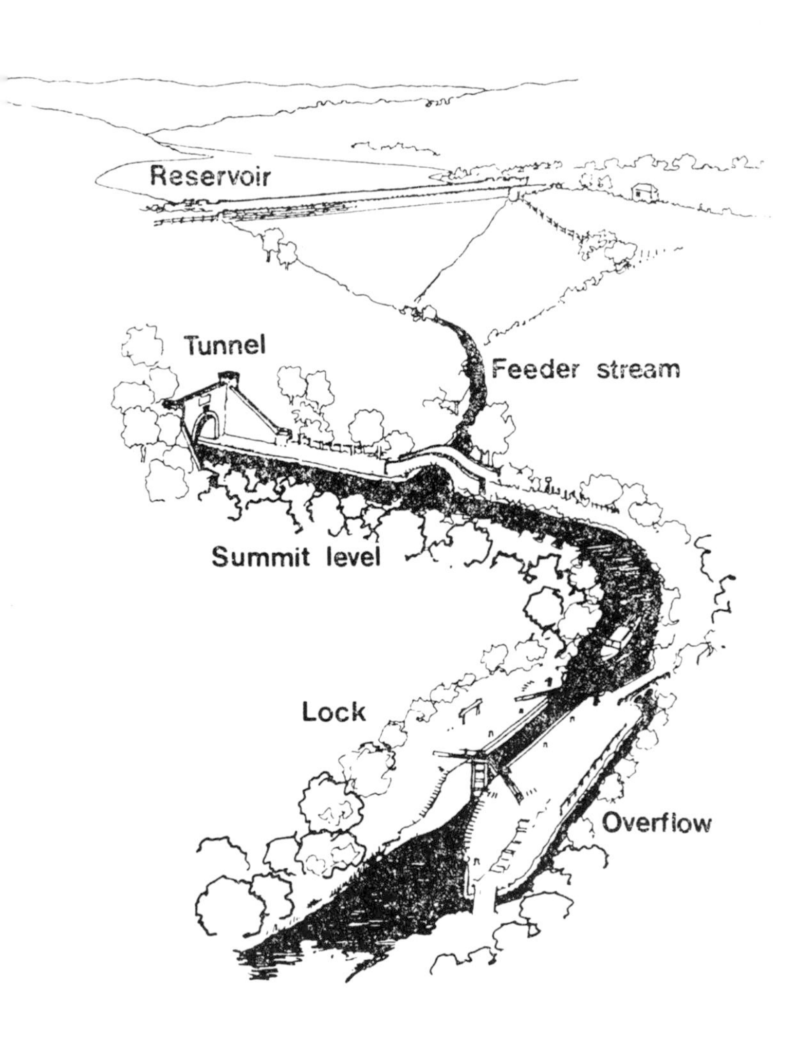

The water supply to a canal.

weight (50.8 kg). So the Duke did obtain the very first Act for an artificial waterway that was completely independent of any river. James Brindley, who was then in his early forties, was employed as the engineer. The first section of the canal was opened in 1761, and it incorporated the first stone-built aqueduct, at Barton, which captured the imagination of the public, as it carried the boats for 200 yards (183 m) at a height of 39 feet (12 m) above the river Irwell.

The water supply for the Bridgewater Canal came from the drainage of the colliery workings, and at the end of the canal it was allowed to flow into the river Medlock. The Bridgewater Canal and the river navigations, with their natural water supply, had few problems about water, except perhaps in times of drought. But the artificial waterways that followed had to incorporate systems of collection, storage and supply of water for use as and when required. These had to be planned and constructed in conjunction with the building of the canals. Reservoirs were the most satisfactory method, although supplies were also pumped from rivers. In all cases the supply of water had to be controlled, with regulated outlets and feeders, which were expensive to build and maintain.

At first, before connections were made with other waterways, the Bridgewater Canal did not have any locks, for the route of the canal had been constructed on a level plain. Most of the canals that followed had to have locks, and in some instances a great many were required to enable the canals to go through hilly and even

The Grand Union Canal by Startops Reservoir, Tring.

mountainous districts. The locks act as steps between the series of levels. Every time a lock is used to lift or lower a vessel to a different level, at least one lock full of water is taken from the pound above, so all the time, with the passage of craft, water flows down the canal from the higher levels. The larger the lock the more water is used, and because of the trouble and expense in obtaining water supplies to replenish the summit pound it has always been essential to conserve supplies as much as possible. Water conservation is one of the reasons why most of the early canals incorporated narrow gauge locks. The canals with wide locks were generally those constructed later and usually only when adequate supplies of water were available. Even then, at times, there were shortages. Unlike the narrow canals, these wide canals, with locks usually able to take a pair of narrow boats side by side, did not form a complete network.

Rivers are natural features and Acts of Parliament were needed to make them navigable and preserve certain rights. Consideration had to be given to existing users and riverside property owners, such as mill owners, to ensure that the water power was not taken from the mills and also that changing water levels, as a result of installing weirs and locks, did not flood private property. New canal projects also required Acts of Parliament before construction could start, but they had to take other factors into consideration. Canal Acts were really compulsory purchase orders that had to take into account natural drainage and the rights of access. This applied particularly to landowners who had to have access to their property once a canal had been cut through their land. They made provision for bridges and, in some cases, decided on which side of the canal the towpath had to be.

Canals were usually constructed because groups of men with foresight and vision saw the advantages that would materialise and so committed themselves to their construction. These groups, mostly consisting of men of influence, such as mayors, aldermen, merchants, lawyers, churchmen and landowners, who realised the potential benefits to be derived from a canal, spent their time and risked their money. They employed a professional engineer to survey the route and prepare plans, and they would provide money themselves or obtain finance for the project, before making formal application to Parliament for an Act. When Parliamentary approval was secured work on the new waterway could proceed. Obtaining an Act was far from easy for the promoters, for they had to prove that the waterway would be a worthwhile proposition and would be of benefit to the area. In addition they had to be prepared to alter plans or even change the route to satisfy objectors, as many landowners did not relish the prospect of a canal cutting through their estates. With work proceeding, the promoters would appoint directors from amongst themselves to take over control but employed staff for the everyday running and operation of the waterway.

As there were no mechanical aids the canals had to be dug entirely by men using picks, shovels and wheelbarrows. The canals ran through ground that consisted of many types of material, ranging from hard rock which would hold water when the canal was finished to sand that allowed water to seep away. To overcome seepage from the canal, clay puddle was used. This was a mixture of common clay and a little water that made a very pliable material which could be placed exactly when and where it was required. A thick layer would act as a waterproof membrane. This clay puddle proved to be the ideal material, for today hundreds of miles of canal are still lined with it. It is still used to stem breaches and repair leaks, although modern materials such as polythene sheet and concrete are also used.

The men who dug the canals were called navvies because most of them had previously worked on the navigations. When working on the new canals they were either employed by the canal company itself or by a contractor who dug part or all of the canal for the company. They worked according to the engineer's instructions, digging the canal with a deep section down the middle to provide a channel for the boats. Besides the navvies there were men of specialised skilled trades, such as masons, joiners and bricklayers, who were kept busy erecting bridges, aqueducts, locks, numerous canalside buildings and tunnels.

The canal age, which started during the reign of George III and gradually ended when Victoria was queen, lasted for about one hundred years. The canals helped to change Britain from an agricultural society to one based on manufactured goods, for they provided a cheap, safe method of transport. As the first canals were so successful, many schemes were put forward that resulted in canals completely criss-crossing the country until in total there were over 4,000 miles (6,400 km) of interconnected canals and navigations. However, there were some failures which never even recouped the money invested in their construction; these were the canals built to serve primarily agricultural areas. But on the whole canals were successful and, with little competition at first, most of them made a great deal of money for all concerned.

Tramroads with horses to haul the wagons were used for many years to transport materials such as coal and stone from mines and quarries to waterside wharves. At first no great future had been considered likely for this form of transport because of the limited capabilities of the power source. Then the success of the early

The track of the Hay Inclined Plane at Ironbridge, Shropshire.

steam engines, which provided a very satisfactory means of propulsion, revolutionised the whole concept of rail transport. From very small beginnings, in only a few years railway lines were to spread throughout Britain, much to the detriment of inland water transport.

Most of the railway companies were not satisfied with a share of the trade nor were they prepared to generate trade that would materialise from the natural expansion of industry and commerce, but they went out to take everything. Many did this either by undercutting rates or by trying to buy control of canals. Most of the canal companies were financed from tolls received and wharfage dues as they did not operate vessels themselves. The offer of guaranteed dividends was very attractive, especially for doing nothing, and so many canals passed into railway control. Other railway companies, intent on extending their lines, bought control of certain canals, so that they could get a bill through Parliament; this method satisfactorily squashed what would have been their chief opponent. All this led to a very unsettled period for water transport, which was experiencing competition for the first time.

When the railways were in control they were not concerned about the canals or the private carriers but did everything possible to discourage water traffic. They put up their tolls to the carriers to such unreasonably high rates that it became impossible for many of them to continue trading. Alternatively they neglected maintenance and repairs and this in turn soon stopped traffic. They nearly succeeded in their attempts to gain a monopoly because with the loss of trade many canals silted up and were eventually closed. On the other hand, a few railways which had taken over a waterways system that extended from their own territory into the area of a rival railway company encouraged trade and this resulted in the establishment of a number of railway transhipment wharves. Fortunately the railways did not have everything their own way, for a few canal companies did resist the railway takeover bids and fought hard against railway competition, with the carriers introducing steam-powered vessels to speed up the movement of goods. These canal companies that continued trading experienced hard times for many years, trying to compete with undercutting by the railways, but some managed until the railway stopped the practice, and then trade levelled out once more.

At the same time some of the large and very prosperous navigation companies were determined to try to compete, so they embarked upon improvements to their networks, by constructing further canal sections to reduce the overall mileage and speed up journey times. They enlarged the locks to enable bigger vessels to operate and also introduced steam-powered vessels, including tugs, which proved very satisfactory on specialised workings. All these ways of enabling them to carry bigger payloads more quickly and profitably made it possible for most of them to compete.

As a result of the expansion of the railway network in the nineteenth century it was no longer practical to build many more canals, although by this time government legislation had finally stopped railway takeovers that encouraged a monopoly and stopped free trading. This government action seriously restrained railway influence over canals and even allowed a few to operate independently once more. The Sheffield and South Yorkshire Navigation was one of the freed waterways and they, together with the Aire and Calder Navigation, built the New Junction Canal. This waterway, 5.5 miles (8.9 km) in length, was opened in 1905 and connected the two large navigations, but it was the last canal to be constructed in Britain.

After 1900 a number of government surveys and reports were undertaken to consider the future potential of the national network of waterways. All were in favour of improvements, with modernisation recommended for most of the remaining waterways to make them more viable in the future. Unfortunately very little action followed and they were left for the most part as they were.

The biggest drawback to the waterway network is that there are locks of many different sizes, with the result that very few vessels are able to travel the whole network. Some locks are long and narrow, others short and wide, and within these limitations they impose severe restrictions, so that a vessel able to travel on the whole network must be no more than 7 feet wide (2.1 m) and about 60 feet long (18.3 m) and so is able to carry less than 20 tons.

3. Canals since the 1920s

In the early 1920s the employees of many canal companies demanded a shorter working week and more pay and as a result a number of companies closed down their carrying concerns. But higher operating costs were not the only problem facing the inland waterways for the activities of a new competitor, the motor lorry, were beginning to be felt. At first the small capacity motor lorries had only a limited effect, but as the years passed their increased efficiency, through carrying on better roads, was to have a marked effect on inland water transport.

Narrowboat operations were gradually reduced owing to the increased costs and competition, but there were some who believed that they had a future. A great boost came with the formation of the Grand Union Canal Company Limited on 1st January 1929, when five separate waterways amalgamated. The new company immediately concentrated upon improving its network and obtained financial aid from the government to modernise the route between London and Birmingham. In 1934 they formed the Grand Union Canal Carrying Company Limited, with plans to succeed where others had failed. During the next few years they ordered many new boats, the largest order being in 1936 for eighty-six new pairs. Despite their enterprise they were not altogether successful, experiencing both financial problems and difficulties in obtaining enough satisfactory crews to keep their large fleet constantly operational. Nevertheless, they continued until after the Second World War and nationalisation, when their fleet formed the basis of the British Transport Commission's narrowboat fleet.

During both the First and Second World Wars all the waterways owned by the railways and the majority of the private ones that still had commercial traffic came under government control. On both occasions they were afterwards handed back to their owners. At the end of the 1939-45 war, because of almost six years of neglect, waterways were generally in a very run-down state. On 1st January 1948 they were nationalised and the government again took over control of the inland waterways. This government-controlled system did not include tidal rivers like the Thames, the Humber and the Mersey or the canals that had not been controlled previously, and this left a number of canals and most rivers independent of the nationalised body.

The British Transport Commission's area of control included the railways, a number of docks and also road haulage. The inland waterways in particular became the direct responsibility of the Docks and Inland Waterways Executive. British Waterways managed the canals and navigations, on which they had several large carrying fleets operating alongside independent carriers. The private operators had to pay tolls to British Waterways, which was in direct competition with them. This practice was weighted against the private firms because, if the nationalised body wanted, it could either reduce its own tolls or alternatively ensure that another section of the Commission, such as the railway, obtained the work instead. At first some good contracts were lost to the waterways because of the internal transfer of information, but this was soon stopped. In 1962 an independent body, the British Waterways Board, was set up to manage the nationalised inland waterways and since shortly after its formation the Board did its utmost to safeguard existing work and secure further contracts, with the carrying operations coming under the Freight Services Division for a while. In the late 1980s British Waterways sold off or closed down its remaining carrying fleet. All transport is now by private operator.

Before the Board took control in the early 1950s, there was still a great deal of long distance narrowboat traffic. Within the Birmingham district, which was the centre of narrowboat operations, there was over a million tons of short haul traffic every year, but by the early 1960s most of this had gone because of the opening of motorways and closure of collieries. Because of continuous problems and because they were no longer profitable on 14th September 1963 the British Waterways Board stopped using its narrowboat fleet for commercial carrying. By this time also most of the large fleet of independently operated narrowboats had been disposed of, many going for scrap, while the others found a new lease of life when converted into pleasure craft. However, a few working narrowboats still remain but are now mostly owned by enthusiasts who try to continue operating them in the traditional manner. This is very difficult, and during the summer months most of them are used by youth organisations for camping holidays. In the winter a few endeavour to find cargoes, and some have been enterprising enough to obtain the occasional load, varying from bicycles to tractors. Some do have regular winter work delivering coal to householders alongside the canals, but this trade, in total, does not exceed 1,000 tons per year.

The last regular long-distance working of the old pattern, using narrowboats, was the movement of lime juice in barrels from Brentford to Boxmoor on the Grand Union Canal, which ceased in 1980. In the Birmingham area a number of boats without cabins, called 'day boats', transported rubbish from factories until the 1980s. On the Caldon Canal, Johnson Brothers (Hanley) Limited, with a number of canalside works along a 4 mile (6.4 km) section of canal, for many years experienced problems with the transport of pottery between their works, so they decided to try the smooth canal. In 1967 they began to operate a new craft, *Milton Maid*, to transport 20 tons of pottery at a time between the works. The waterway operation proved to be a success and as a result in 1973 they put a second vessel, *Milton Queen*, into service. Rationalisation of factories, however, brought this traffic to an end in 1995.

During the early 1960s most people found themselves with more spare time and more money to spend, and many were able to devote some of this time and money to leisure pursuits, especially outdoor activities. This brought about an appreciation of the canals and rivers, which provided people with the opportunity to escape from the pressures of modern living. From then on more and more people explored the waterways, by foot along the towpaths or by boat, so that cruising became very popular. This continual increase in the number of private boats and of holiday craft for hire resulted in the canals becoming busier and an increase in the facilities provided. Some local authorities realised the environmental and amenity value of canals. Canal societies and voluntary restoration groups were formed and with the aid of boat rallies and other public activities they publicised the waterways and their many aspects, encouraging still more people to take an interest. As more people found satisfaction and enjoyment from waterways, concern grew about the future of one of Britain's national heritages, closure of canals stopped, and a great increase in the number of restoration projects followed. This has resulted in recent years in the reopening of a number of former derelict canals, to provide more open waterways for the ever increasing numbers that come for the first time every year to sample and enjoy inland waterways.

In 1968 a report from a survey taken on the future of the waterways showed that most of the inland waterways were used only for pleasure, but that a few wide waterways, especially those close to river estuaries, still had a substantial amount of commercial traffic. The increased use of the waterways for pleasure and the public concern for their future resulted in an Act of Parliament. This made the Board's waterways available for use by the public in the present and in the future. Certain canals were designated as 'cruiseways', available for cruising and recreational use. The waterways used commercially were classed as 'commercial', while the others, which had at that time little or no use, were the 'remainder'. Since then though, because of the work by volunteers, local councils and the Board, many canals in this last group are busy and active for recreational purposes.

The commercial waterways of the British Waterways Board total 336 miles (540 km) and include, in the north-east, the Aire and Calder Navigation, the New Junction Canal, the Calder and Hebble Navigation (Wakefield to Dewsbury), the Sheffield and South Yorkshire Navigation (Keadby to Rotherham) and the Trent Navigation (Gainsborough to Nottingham), the Weaver Navigation in the north-west, the river Severn downstream of Stourport and the Gloucester and Sharpness Canal in the south-west, and the River Lee Navigation (Hertford to the river Thames at Limehouse and to the tail of Bow Locks) in the south-east. In Scotland there are the Caledonian and Crinan Canals. Not all these waterways have regular daily commercial traffic, but some of them do, while other waterways with commercial traffic, ranging from the river Hull in Humberside to the river Yare in Norfolk, do not come under the authority of the British Waterways Board.

The cargoes now carried on the Board's commercial waterways, which now includes the river Ouse upstream of Goole, totals over four million tonnes annually. Nevertheless the future for most of them as viable commercial waterways is not very promising; this is despite considerable investment in some inland navigations in the north east. During 1978 a series of improvements was completed on the Aire and Calder Navigation to enable 700 tons capacity vessels to reach Leeds from the tideway. In the same year, after many years of pressure, money was allocated to modernise the commercial section of the Sheffield and South Yorkshire Navigation. Following the completion of this major improvement scheme on 1st June 1983 traffic figures have been disappointing. This is primarily due to recessions in trade with the decimation of both the steel and coal industries in the area and the closure of small electric generating power stations.

While the Aire and Calder Navigation does still retain some considerable traffic, it is mostly on short term contracts, operated by large capacity craft. On the whole the future success of these two connecting British Waterways Board's improved waterways in particular may depend upon current concern for the environment and energy conservation. For it is a fact that large inland waterway craft are five times more economical in the use of fuel than heavy lorries and their operation relieves the pressure on overcrowded roads. There is also the added bonus that inland waterway transport is much safer, does not result in damage to the fabric of buildings and has much lower noise levels.

4. Architectural features

When work was under way on the new canals, especially in winter, they were an almost continuous sea of mud and this made a scar on the landscape that took many years to heal. Now they blend into the landscape. The watercourse is fringed with green, and the hedges that border many miles of towpath provide homes for countless birds and other wild life. Practically every type of British wild flower is to be found somewhere along their banks, with the colours enhancing the beauty that nature has bestowed upon the waterways.

Building a canal was an enormous undertaking necessitating earthworks of huge proportions and on such a vast scale that thousands of men were kept at work often for many years. Canal construction created the profession of civil engineer, and one of the first and most renowned was James Brindley. His success with the Bridgewater Canal brought him fame, and he soon found himself in big demand and occupied in a considerable number of large projects, such as the Trent and Mersey Canal and the Staffordshire and Worcester Canal.

The engineers had to use all their initiative, skill and ingenuity to succeed. Not only was the work they were doing new but before 1801 there were no Ordnance Survey maps and very few other maps that could be relied on. So they had either to make their own surveys or to use locally made, often inaccurate maps. The result was that if they failed to detect differences in geological formations they could have constructional problems that increased costs considerably over the estimates.

The canal engineers came from many different backgrounds. Amongst the most famous, Brindley had formerly been a millwright at Leek in Staffordshire, while Telford had been a mason. Smeaton was the son of an attorney, while Rennie was the son of a farmer. Many learned their profession from working alongside the first acknowledged experts, such as Jessop, who had worked with Smeaton, and both Robert Whitworth and Thomas Dadford, who achieved great things after working with Brindley.

An eminent engineer with overall responsibility for the building of a canal had resident engineers to supervise day to day problems. Many of the site engineers did more than that, for certain company offices, aqueducts and tunnels were designed by them. The contractors regularly designed and built in a simple manner many of the various canalside buildings such as lock keepers' cottages, toll offices, stables and small workshops.

Fortunately, because so many waterside buildings have remained intact, they now illustrate civil engineering techniques and design

One of the many swing bridges on the Leeds and Liverpool Canal.

from the eighteenth century. They provide the opportunity to study the changes from early basic techniques to the later sophisticated structures of Rennie and Telford. Most of the early canal buildings were purely functional, of a pleasing simplicity that gives the impression that they were built without any conscious design. Later, the nineteenth century produced some very decorative structures with elements of neo-classical design or, in certain areas, simulated Gothic.

The early canals clung to the contours of the land, meandering through the countryside in an attempt to avoid as many changes in level as possible, the gentle bends forming part of the landscape. They required fewer locks, which were expensive to build, while the long pounds provided an extra reservoir for the precious commodity, water. A single lock that could accommodate one narrowboat also used less water. As the years passed, building techniques improved and the second generation of engineers, like Rennie and Telford, were able to make full use of the experience gained. They were intent on creating the shortest possible routes and built straight canals that overcame the natural obstacles in the way. One technique that was evolved by the second generation of canal engineers was cut and fill. The material excavated in one place to make a cutting was used a little further along to level out hollows and make embankments. Attempts to speed up water transport with direct routes required more aqueducts and long flights of locks so they created elaborate structures to enable the canals to go through deep cuttings and high over embankments and this showed their engineering skill. During this period of canal construction some very substantial wide canals were built and in addition attempts were made to straighten out some of the old contour routes of Brindley and his counterparts, such as the Birmingham Canals main line and the Northern Oxford Canal. The new sections of canal bypassed old loops, most of which then served no further purpose and disappeared, although one interesting old loop of the Oxford Canal remains, running through a disused tunnel close by the church at Newbold on Avon, near Rugby.

Bridges

When the canals were cut across the country they disrupted many existing roads and rights of way. These had to be reinstated so many bridges of various sizes had to be built. The most common type was the humped back variety, many of which were built to enable a farmer to move his livestock from one side of the canal to the other. They were built of the local brick or stone,

A split bridge at Lapworth on the Stratford-on-Avon Canal.

usually to a company standard design. Some canal companies erected cheap movable types that either lifted or swung sidewards to allow passage along the canal. Often, to appease the rich landowners who lived in nearby stately homes, ornate bridges were built to carry approach roads over the canal, but especially in towns and cities large robust bridges were built to carry highways and to make a good impression on the local residents. Also over the canal, especially at junctions, bridges were built to enable horses to gain access to the towpaths. These bridges of graceful lines were known as 'turn-over', 'roving' and 'snake' bridges, names which describe the overall shape. Most of them allowed the horses to proceed from one towpath to another without the need to disconnect the tow rope.

Many fascinating bridges were built near locks. Mostly footbridges, they were provided for the use of lock operators. Some very elaborate types were built, including cantilevered bridges with a split down the centre which allowed the tow rope to be dropped down the resulting gap and so speeded up the departure of the boat from the lock, and some very simple ones which were nothing better than a fixed plank, some even without handrails.

Later still, using mostly traditional building materials, bridges were built with wider curved arches, elegant parapets and ornate balustrades. This second generation of canal bridges was built with full consideration for line and proportion, for by then it was felt that it was not sufficient for a bridge to be purely functional but that it had to have a dignified appearance.

In the eighteenth century cast iron supports were used in structures, mainly as invisible girders and stanchions. From the nineteenth century decorated cast iron was used to advantage increasingly in canal bridge construction, and some of the resulting works are masterpieces. Horseley Ironworks in particular produced some very beautiful bridges which consisted of sections bolted together, and these were erected throughout the Midlands. Some of these bridges have a graceful span and are topped with a continuous handrail that sweeps over the canal. They can be seen in the Birmingham area and on the Northern Oxford, carrying the towpath.

Later the canals were to be crossed by numerous bridges, and many fine examples of railway architecture can be viewed to advantage from the waterways. Modern roads and motorways demand a bridge that crosses over the waterway in a sweeping arc or at a specific level well above the waterway. Many of these bridges, of steel and box construction as well as others of prestressed and reinforced concrete, are hardly ever seen by those they were built to serve but can be appreciated from the canal.

Tunnels

The early engineers preferred single bore tunnels. They did not like deep cuttings, for the problem of slipping sides was usually insurmountable. When work started on a tunnel it was usually at both ends simultaneously, and digging also proceeded on the route of the tunnel from the bottom of shafts down which men were lowered. These shafts, linked together at a given level, provided the route for the tunnel, but the use of many different digging parties, working towards each other, often resulted in far from straight tunnels. Many of the shafts were later lined with brick and used to ventilate the tunnels, especially after the introduction of steam-powered tugs to haul vessels.

These subterranean watercourses, often lined with brick or stone, are basically horseshoe-shaped in section. Some of them extend great distances and so took a very long time to build. The earlier ones had no towing paths, so the horses went over the top to meet the boats at the far end, while the boats were 'legged' through by men lying on their backs and walking along the sides of the tunnel.

Improved building techniques resulted in wider tunnels that allowed two boats to pass as well as the provision of a towing path. But even so many horses would refuse to enter the tunnel and had to be taken over the top of the hill. Tunnel entrances are mostly unobtrusive, but there are a few exceptions with very fine portals.

Shortwood Tunnel on the Worcester and Birmingham Canal.

The Almond Aqueduct on the Union Canal in Scotland.

Aqueducts

The new canals had to cross over many existing rivers and streams and over roads and valleys, so aqueducts were built. The early ones, as constructed by Brindley, were never built to great heights. They were made entirely of masonry, with the channel lined with puddle, and were of massive dimensions. Aqueducts with watertight cast iron troughs, which were bolted together in sections, were to follow, especially in ironmaking districts. These became the most common type and were supported at each end by a strong masonry abutment.

Many of the short aqueducts that were required to cross over roads have very pleasing lines. The Stretton aqueduct, which carried the Shropshire Union Canal over the Holyhead road, has an artistic quality. Longer aqueducts crossing valleys, such as Chirk on the Llangollen Canal, have stone piers to support the trough, which in this case is encased in masonry. The Pontcysyllte aqueduct, on the same canal, does not have its trough concealed. Completed in 1805, this is a masterpiece by Jessop and Telford. It has a trough 1,007 feet long (307 m) and dominates the landscape.

All-metal aqueducts, such as the one designed by Telford to carry the 'Engine Arm' over the Birmingham Canal Navigations' main line, are wonderful examples of engineering techniques. Here the cast iron trough is mostly concealed by numerous delicate

Gothic arches. The use of cast iron for decoration gave plenty of scope, for some aqueducts have some fine decorated balustrades. The old Stanley Ferry aqueduct on the Aire and Calder Navigation, opened in 1839, is of unusual design. It has two great bow-shaped arcs that support the trough on either side, whilst the ends are supported on stone abutments.

Most of the people who cruise along the waterways do not notice the dozens of aqueducts, all of which are worthy of a little attention. They should stop for a few minutes to have a look, if possible from a lower level, so that they can appreciate the lines created by the engineers.

Canalside buildings

At first the canal companies did not consider employing engineers to design workers' homes, such as lock and bridge keepers' cottages. They were usually built of the same materials and to the same plan as the local cottages, perhaps the only noticeable differences being a side window to give a view of the canal. Afterwards they were built to the specific designs of individual canal companies, some of them even being influenced by Grecian architecture. The second generation of cottages on certain waterways were provided with observation porches or large bay windows.

Telford's aqueduct carrying the Engine Arm of the Birmingham Canal Navigations over his new main line.

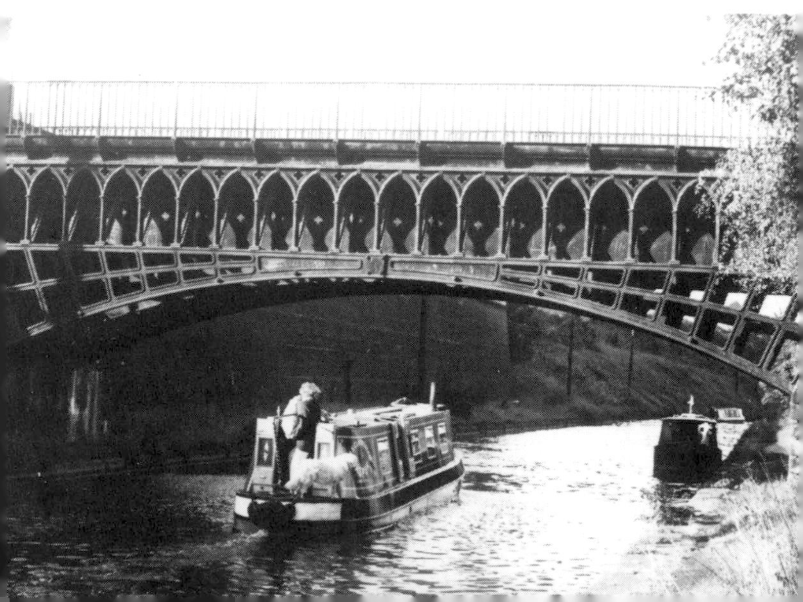

The stables on the Delph flight of locks on the Birmingham Canal Navigations.

As arteries of transport the canals generated trade and, to cater for this, numerous warehouses and other waterside buildings for industry and commerce were continually needed. The expansion of trade by water resulted in the creation of new towns, built at waterway junctions where storage, transhipment and distribution boomed. Along with the normal business of the canal companies, which required office accommodation and houses for the workers, associated trades such as boatbuilding flourished. Two such towns that have their origins in early canal development are Goole on the Aire and Calder Navigation and Stourport on the Staffordshire and Worcester Canal.

Warehouses of every description, both large and small, were built of stone, brick and timber. Many still remain today, over a hundred years after they were built, as strong as they ever were, and usually positioned alongside a former busy canal terminal or basin. They had oil and later gas lighting to allow for the loading of vessels throughout the night. Many large hand-operated cranes were used at wharves and warehouses, along with hoists to speed up the movement of goods. Much of the early mechanical handling equipment has gone, but fortunately a good number of excellent examples remain, most of which have been restored to working order in recent years.

To house the thousands of horses that for nearly two centuries trod the towpaths there were many hundreds of stables, some with haylofts above. In recent years many of these stables have been converted to house canal shops and other services for the growing number of users of the waterways. Small buildings were also built to house a blacksmith shop and other canal trades such as rope and sail makers. The canal companies also built repair yards and maintenance depots, where they made new lock gates and the fixtures and fittings for them. They provided many other pieces of equipment needed to keep the waterways in a good state or repair and also specialised maintenance vessels. The large depots acted as servicing hubs for the smaller yards, which were responsible for everyday care and maintenance. Many of these yards with their numerous interesting buildings remain, still carrying out the functions for which they were originally built.

Alongside the canals are a multitude of small fixtures and fittings, from mooring rings to bollards, some of them extremely worn, with very clear rope-marked grooves. Fortunately many waterway undertakings made a practice of numbering and dating things. So signs, markers, boundary posts, number plates and so on abound, located on bridges and buildings, locks and aqueducts, tunnels and warehouse walls. Some of them provide a wealth of information but all are worthy of a second glance.

The achievements of the engineers and the workers are spectacular, especially as most canals were constructed in an age that relied entirely on muscle power.

The Etruria workshops face the statue of James Brindley across the Caldon Canal.

5. Locks

A canal lock is a fascinating and very satisfying piece of mechanism to operate. Simple in design, it was the first major development that made inland water transport possible. To overcome difference in water levels, lifts and inclined planes have been tried and, for the most part, discarded and now over four thousand locks exist in Britain, varying in size from some that are able to pass sea-going vessels to the smallest, the narrow locks. Narrow locks are able to accommodate a boat about 70 feet long (21 m) with a 7 foot beam (2.1 m). Most are fitted with a single gate at the top end, that is the upper pound end, and have a pair of mitred gates at the bottom end. The rise and fall varies from a little over 14 feet (4.3 m) to only a few inches at a stop lock. These are usually to be found where two separate canals met and were originally built to ensure that the water in one canal was not taken by the neighbouring one. The paddle gearing fitted to narrow locks is not standard throughout the system and many types are to be found, for the old companies had different ideas about which were the most efficient and reliable. Basically locks are provided with both ground and gate paddles, but there is a great variety of machinery, with some very ingenious systems of gearing, most of which incorporates a catch or pawl to hold the ratchet once the paddle has been wound up.

On certain canals and on river navigations, where usually there was no undue water shortage, larger locks were constructed to allow barges to pass. On the navigations the locks were initially built to accommodate the barges already in use in the local areas, most of which were about 55 feet (16.8 m) long and 14 feet (4.3 m) wide, but over the years many of the navigations enlarged their locks so that larger vessels could operate.

The gates on wide locks are positioned in pairs because the water pressure would be too great for a single gate or it would be too heavy for manual operation. On a number of waterways, such as the river Nene, certain locks have had the bottom gates replaced with metal guillotine gates. As with the locks found on the narrow canals, many different types of paddle gearing exist. A number of waterways have been modernised in recent years and these have electrically operated locks, the gates and paddle gearing being worked by a lock keeper in a control cabin situated above ground level. At such locks it is the usual practice to have traffic flow controlled by traffic lights.

At a manually operated lock each gate is fitted with a long projecting balance beam. This provides leverage to open the gate, and this is best done, whenever possible, by pushing backwards

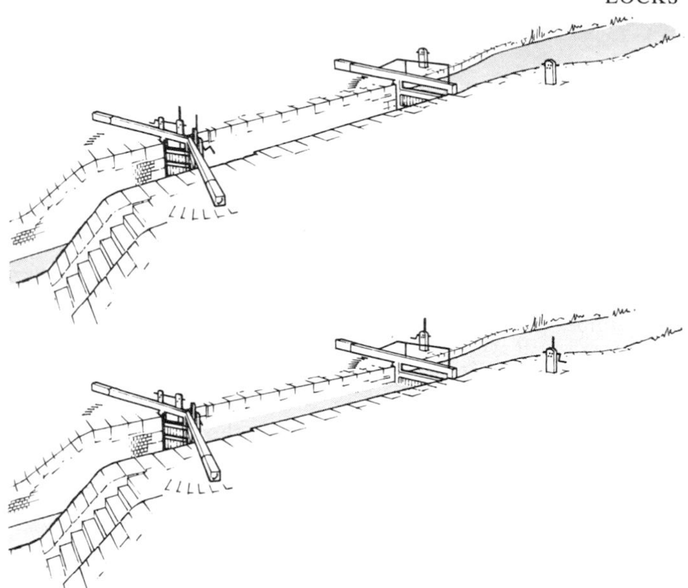

Lock operation. In the upper drawing the top paddles of the lock are open and the bottom paddles are closed in order to fill the lock. In the lower drawing the top paddles are closed and the bottom paddles are open, to empty the lock.

against this beam. To ease the closing of an open gate a hand grip is usually provided at the end of the beam, but at locks with gates which are particularly heavy or difficult to move a winch or wheel operated gearing is often provided. Locks with guillotine gates moving in a steel framework are operated electrically or by numerous turns of the windlass. Locks usually have rectangular chambers of brick, stone, concrete, iron plates or steel piling and there are even a few with turf sides in use. A few locks, such as the Pershore Lock on the Lower Avon, also have a diamond-shaped chamber.

A slight change in level was overcome with a single lock, while a greater change required a number, known as a flight of locks, with each lock separated by a short intervening pound. There are numerous examples of flights of locks throughout Britain, the most notable one being on the Worcester and Birmingham Canal at Tardebigge, where there are thirty narrow locks. At Devizes on the Kennet and Avon Canal the Caen Hill flight consists of twenty-nine wide locks. In other places a number of locks are fastened together, sharing intermediate gates and they save on the use of water and effort. The Leeds and Liverpool Canal has some fine examples, with the Bingley Five Rise the most spectacular. In Scotland the Caledonian Canal has three very impressive sets of staircase locks, the best known being the eight locks at Banavie.

The key to the inland waterways is the windlass, a cranked handle that fits either the 1 inch (25 mm) or 1.25 inch (32 mm) square spindle, or both, which are found on the majority of paddle gearing. The lock is like a box, consisting of a chamber with gates at both ends. At each end of the chamber is a sill against which the bottoms of the gates close. In the case of a pair of gates, they close together in a slight vee, against the upper pound and the pressure of water, which keeps them closed. The pairs of gates are mitred together to withstand the pressures and so make a watertight seal. The flow of water in and out of the lock is controlled by paddle gearing, either gate or ground paddles, or both. The ground paddle gear is on the lock side and moves a wooden door below a ground level that acts as a shutter over an aperture at the end of a culvert that leads into the lock chamber. The gearing on a gate moves a similar wooden door that covers a small opening in the lower section of the gate. The gates can only be opened when the level of water at both sides is equal. For a boat to descend a lock that is empty, the lock must first be filled. This is done by opening the ground paddles and then those on the gate with a windlass, after first making sure that the bottom gates and paddles are closed. When the water in the lock is the same level as the upper pound, the top gates are then opened to allow the boat to enter the lock, then both the gates and the paddles are closed. The paddles at the bottom end of the lock are then opened to allow the water to flow out of the lock into the pound beyond. The boat lowers in the lock as the water level drops, until the water level in the lock and that in the lower pound are equal.

The bottom gates can then be opened to allow the boat to continue on its way. Before the crew leave the lock they must ensure that all the gates and paddles are closed. When a boat is descending a lock there is the possibility of the stern section of the boat catching on the sill of the top gates, so one must always make sure that the stern does not ride down too close to the top gates.

When travelling upstream the procedure is reversed. Water entering a lock too quickly can knock a boat about and damage it, so the boat should be secured to the bollards provided. This is especially true with a narrow-beamed vessel in a wide lock. Also, with a vessel with an open cockpit at the fore end, there is a possibility of sinking it, especially if water is allowed to flow uncontrolled through the gate paddle openings. When ascending a

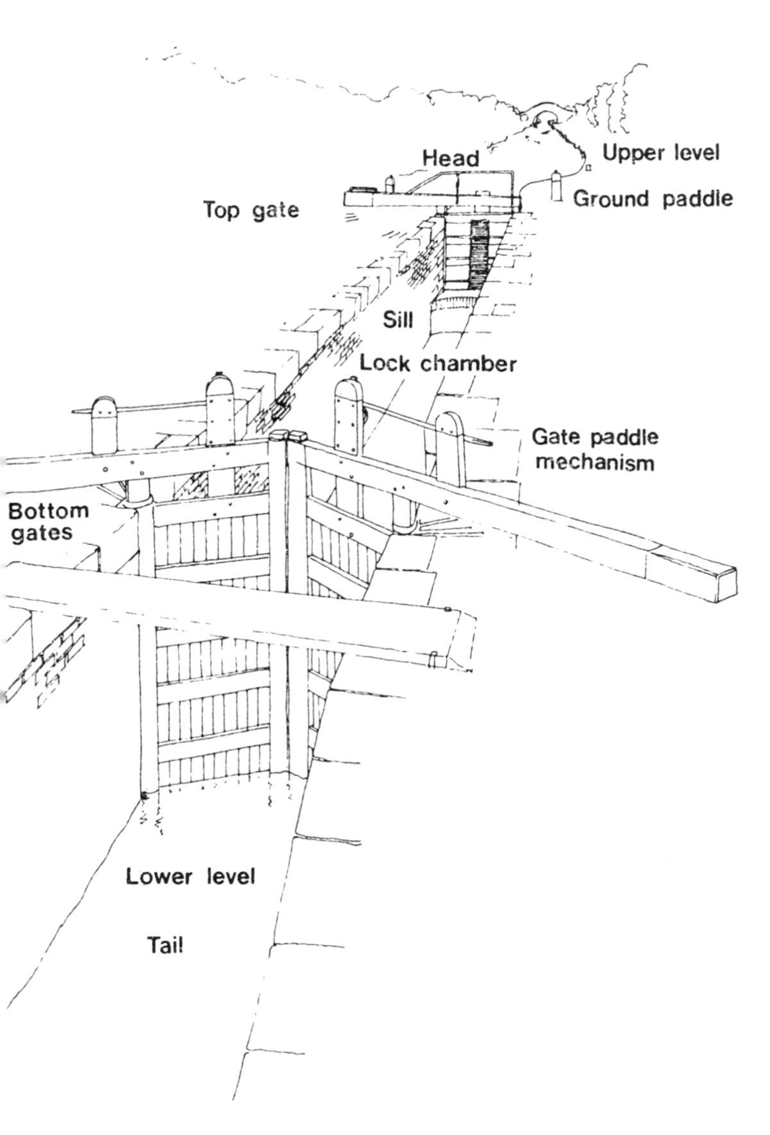

The fixtures and fittings of a typical narrow lock.

lock the ground paddles are used first, followed when safe to do so by the gate paddles. Always take care, watching every development, and do not rush. Remember that a paddle opened on the top right-hand side allows water to enter the lock and flow against the left-hand side, from where it is usually redirected to the right-hand side. This can be used to advantage by holding a narrow-beamed vessel reasonably safe for a while against the right-hand wall of a wide lock. In certain narrow locks, with a vessel sitting halfway along the chamber away from the gates, the water entering the lock will flow under the boat to build up against the bottom gates for a short while; then it flows forward, sometimes taking the boat with it, to smash against the top gates. This is not a regular action, but it can occur from time to time. There is also the slim possibility when ascending a lock that the bows of a vessel may catch on an obstruction on the top gates, so do watch all the time that no harm comes to either the boat or the lock gates. Windlasses should never be left slotted on the paddle gearing spindle, as the ratchet pawls do slip at times. Then the windlass may fly off to hit someone or to

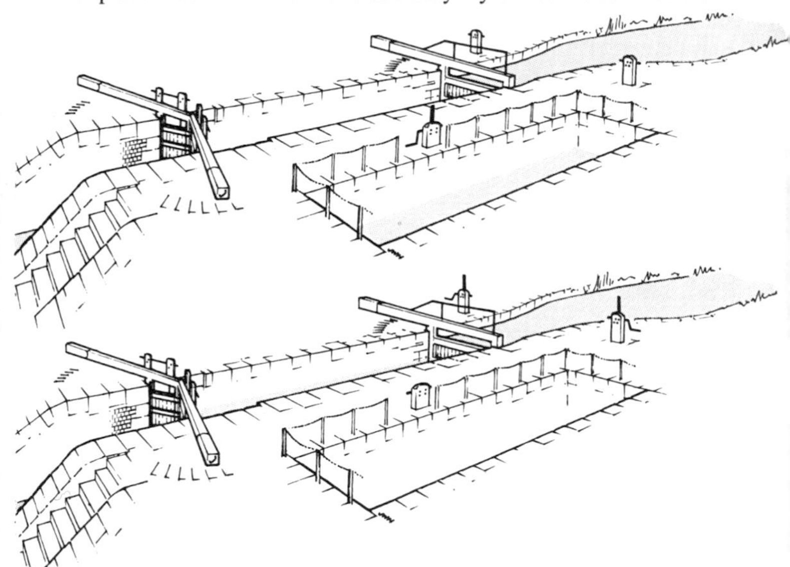

Lock operation with a side pond. To fill the lock, the side pond paddle is opened first (upper drawing), allowing the water to fill the lock almost half full. The filling is completed (lower drawing) by the usual working procedure, with the side pond paddle closed and the water entering the lock from the upper pound.

A view up the flight of locks on the Grand Union Canal at Foxton in Leicestershire.

Batchworth Lock on the Grand Union Canal at Rickmansworth in Hertfordshire.

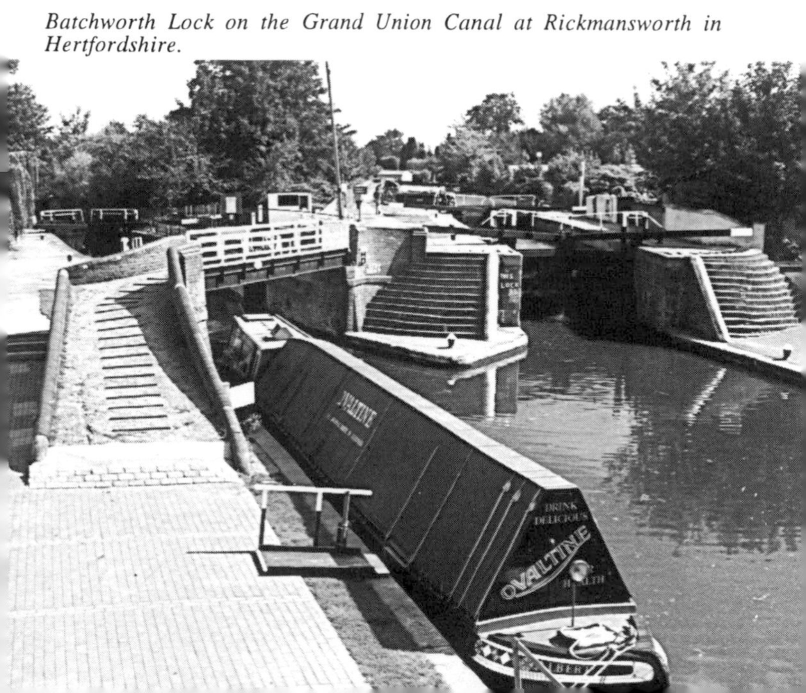

A diamond-shaped lock on the Lower Avon Navigation.

fall into the lock. Never tie a boat to the bollards when descending a lock: let the crew hold the ropes either on the lockside or on the boat by letting the ropes ride around the bollards. In wet weather lock sides and gate footboards can be extremely slippery, so be warned and take care. Lock working techniques are soon learned but, initially, if in doubt watch an expert and never be afraid to ask questions. It is all a matter of commonsense, but think before acting and do try to conserve water.

Water is a precious commodity, and it is essential not to waste it. With this in mind, several canal companies introduced side ponds. Their use enables half a lock full of water to be used again. They are positioned alongside a number of locks on certain canals, and they consist of a brick-lined chamber that is connected to the lock via a culvert and a paddle. To operate when descending a lock, the side point paddle is opened, or, as some say, drawn, first. This allows half of the contents of the lock to flow into the side pond. When the levels of the water in the lock and side pond are equal, the paddle is then closed. The rest of the water in the chamber is then allowed to flow through the lower gate paddles to the pound below. When ascending the lock, the side pond paddle is opened first, allowing the water to fill the lock almost half full, then it is closed. The filling of the lock is then completed by the usual working procedure, using water from the upper pound.

6. Vessels of the waterways

Narrowboats

The forerunners of the traditional gaily painted narrowboats were the starvationers. These earliest of canal boats were simple lean-looking double-ended craft that carried coal in the Duke of Bridgewater's flooded mines. It is said that when Brindley was building the canals he even designed the locks to accommodate this type of vessel. The journeys of the earliest canal craft were short and there was no need for living accommodation, but as the network extended so did the journeys, and a temporary living cabin was fitted. Until the coming of the railways the boats were crewed mostly by men and boys. Then, as a result of the competition and a need to operate at lower rates, the boat owners had to cut costs, so most of them gave up their homes and brought their wives and families to live on board to assist with the running of the boat.

With a family living aboard, the space in the cabins had to be fully utilised and the resulting design was to remain practically unchanged for over one hundred years. Entry into the 10 feet long (3 m) cabin was through two small doors that opened outwards to show the footboards on which the steerer stood. A 2 feet (0.6 m) square opening in the roof above could be closed by a sliding hatch. The footboards were positioned several feet above the cabin floor, with a box that held coal for the range acting as a step. Along the right-hand side of the cabin was a bench with drawers and a locker incorporated. This bench was used as a seat during the day and a bed for a couple of children at night. On the left, located on a low stand, was the coal-fired range and oven, alongside which was a slim floor-to-ceiling cupboard with a door that hinged downwards to make a table. In this cupboard, which was also fitted with two drawers, was stored food, crockery and cutlery. Next was a much wider cupboard in which a double bed mattress and blankets were stored. The main door of this cupboard also hinged downwards, bridging the central aisle, and provided a bed for the boatman and his wife at night. This part of the cabin was concealed from the rest by a lace-edged curtain.

The cabin interior was decorated with wood graining, with panels of painted roses and castles on the bulkheads. White lace-edged curtains, a number of hanging coloured china plates with lace edges and a good deal of polished brass, including an oil lamp, completed the scene. Formerly these vessels were drab looking and it was the woman's influence that brought the adornments, both in the cabin and on the exterior. On the cabin sides were painted the owner's name and base, beside which was a panel that included a painted castle scene. The cabin roof was grained except for a central row of brightly painted diamonds. A detachable black painted chimney was adorned with a number of horizontal brass strips. A water can, known as a Buckby can because the first ones came from Buckby, and a round bowl, fitted with a handle called a dipper, were both painted with the usual rose decoration. Other loose deck fittings included a mop, its handle painted in stripes similar to a barber's pole, a boat shaft and some cotton rope. Many parts of the boat were painted in geometric designs along with moons and playing card symbols. Green, red, blue and yellow were favourite colours.

The horse-drawn boat had a pointed stern, similar to the bows of a ship, and on this hung a large wooden rudder, the main part of which is called a ram's head. This too was well painted and decorated with circles and geometric designs and adorned with plaited ropework and sometimes a horse's tail. The tiller, which curved down to the cabin doors when in use, was reversed so that it pointed upwards and away from the cabin entrance when the boat was at rest.

Until the middle of the nineteenth century the boats worked individually and were hauled by a horse, donkey or mule from the towing path. Then some vessels fitted with steam engines commenced work, some on express duties, while others were used to haul a number of boats. They had their disadvantages, for the boiler, engine and fuel took up much valuable space which formerly held cargo. In addition they required the services of a full-time crew of three men.

After the turn of the century development of the internal combustion engine resulted in a single cylinder diesel engine, a Bolinder, being installed in some boats. As it took up very little space and did not require extra crew it became very popular. The motor boat as we know it had evolved by the 1920s, with the engine housed in an extension of the cabin forward of the living quarters. Access to the 'engine 'ole' was by small doors on either side of the cabin. The bows of the motor boat are the same as on the horsedrawn vessel, but the former has a longer cabin and different aft section. The motor boat has a semicircular stern deck or counter instead of a small cockpit as on the horse-drawn boats, while the rudder is almost out of sight below the water and is operated by a tiller made of round sectional steel shaped like a Z.

As time passed it became the practice for a motor boat to tow a former horse boat that became known as a butty (a buddy); motor boat and butty together were known as a pair. Usually they were operated by one family and the boatman would steer and work the motor boat and the wife the butty, until an older child could take over. The first boats were made of timber, oak and elm being

Narrowboat operations in the early 1930s at Knighton on the Shropshire Union Canal.

At the end of an era, life on the cut in the 1950s.

favoured for their special properties. The practice of building with wood continued at some yards until after the Second World War. Iron and then steel became the more usual building materials during the twentieth century. Most of the boats were of composite construction, that is with a hull of metal and a flat wooden bottom. They were built at numerous yards throughout the system, mostly family businesses that launched only one or two boats a year, in addition to offering dry docking and repair services. The vessels were built to strict dimensions and an accepted standard design although each yard's products could be identified by the discerning eye because of one or two variations. A few large building yards did exist and built substantial numbers of certain classes in later years. The last commercial narrowboats of conventional design built for general cargo work were the 'Admiral' class. These were built by Isaac Pimblott Limited of Northwich in 1960-1 for British Waterways. The study of narrowboats, the different types, their operations and decorations can be most absorbing, and, as much has been written about them there is plenty to read on the subject before going to look at them.

Barges

On the estuaries and inland along the river navigations sailed the barges of regional design. These, tried and tested over the years by

A restored former Grand Union boat at Farmer's Bridge, Birmingham.

wind and tide, had a proven cargo carrying record. As the navigations extended further inland, locks were built to accommodate these vessels and later, when wide-beamed canals were constructed, adaptations of the regional sailing barges operated on them. All the sailing barges in common use were over 11 feet wide (3.4 m), and included the Mersey and Weaver flats, Severn trows, Yorkshire keels and sloops and Thames sailing barges, and on the rivers in East Anglia were the Norfolk wherries and Stour lighters.

Yorkshire keels, originally about 55 feet (16.8 m) in length, had a very large operating area, working on the rivers in the county and also nearby Lincolnshire and up the Trent to Nottingham, with adaptations going further upstream to Burton. Later, after the opening of the wide-beamed Rochdale Canal and the Leeds and Liverpool Canal, they were able to work all the way from Hull to Liverpool. The smallest Mersey flat was several feet longer than the keel, and so could only travel east out of Lancashire to the end of the Rochdale Canal at Sowerby Bridge, because the Yorkshire locks were too short for them.

The sailing barges used their sails whenever possible, but inland where it was not practical because of the numerous locks and bridges they then relied upon horses, but when steam tugs were introduced these were used on many waterways to haul barges which discarded their sails. A number of regional barges were fitted with steam engines and, although not generally accepted, in some areas they became very popular, for instance in Scotland. When internal combustion engines became common they were fitted in what had, until then, been the aft living cabin. The resulting motor barges proved to be very successful and they were built in increasing numbers, although craft without any sails or power, dumb vessels, continued in use until well after the Second World War, relying on tugs and horses to haul them.

For many years the barges were constructed of oak with elm bottoms. They were strong and solid, and they needed to be, for, with a cargo of up to a hundred tons, they were subjected to stresses and strains on the estuaries and to knocks and scrapes, especially at locks on the inland waterways. Wooden vessels continued in service until quite recently. The last wooden motorised barge to be built was launched at Mirfield, on the Calder and Hebble Navigation, as late as 1955. Vessels built of iron were tried in the early nineteenth century, and many were to operate under sail. Barges of riveted construction were popular in some areas, while in others the boatmen remained loyal to wood. In the twentieth century vessels of steel became common. Those built after the 1940s were of all welded construction and with fewer curves than their predecessors, setting the pattern for the modern

'St Audrey', a Sheffield-sized timber-constructed Yorkshire keel, moored and fully rigged. The features of the vessel from the top sail to the leeboard can be clearly seen. No fenders are visible, not even one to protect the stem at the front, so the vessel had probably just been launched and had not yet carried its first load when photographed.

barges now in use.

Barges have carried every conceivable kind of cargo in their holds, which were usually covered with wooden hatches. For many years bulk liquids were transported in barrels, then lead tanks were installed in a number of vessels, but early in the twentieth century the building of bulk liquid tankers as a specialised type of craft became common. These were the forerunners of the modern large capacity tankers, some of which are capable of transporting 700 tons of bulk liquid on certain waterways. Developments have taken place too in the transport of dry cargo. A few modern barges now have mechanical handling equipment installed that includes conveyors and augers and makes them independent of land-based equipment and also speeds up discharging and turnround times.

The sailing barges were built with two living cabins below decks, one fore and one aft. They were made snug and cosy, warmed by a coal-fired range and lit by polished copper paraffin lamps. The sleeping accommodation was provided in bed spaces, which were large cupboards shut off from the cabin by large glass doors. They provided mobile homes for the families, who regularly worked long hours, winter and summer alike. They continued as family boats in many areas until the late 1940s but after that, where barges have continued in service, they have been manned only by men. The living cabin has not disappeared from the modern vessel, although only a few are used regularly, the men preferring whenever possible to go home in the evening. The modern barge has a cabin fitted with all modern conveniences, ranging from electric lighting to refrigerators, as becomes a vessel which carries radar in the wheelhouse. Nowadays the fitting of radar equipment is common for vessels that regularly work on tidal rivers. Modern barges working on the inland waterways have to be economically viable, and so are mostly of large capacity, crewed by up to three men. These craft are expensive to build and cost over half a million pounds each. The emphasis on large capacity craft does not mean that all the smaller craft have been scrapped. Many of these older vessels still exist, some at work, but a lot have been sold to be used as houseboats and the like.

Containers and compartment boats

James Brindley designed the first canal containers, used to transport coal on the Bridgewater Canal. Similar ones continued in use until recent years on the Leigh Branch of the Leeds and Liverpool Canal and the Manchester Bolton and Bury Canal. The containers, made of wood, measured 6 feet by 4 feet 6 inches by 4 feet deep (1.8 m by 1.4 m by 1.2 m), held 35 hundredweight (1,778

A timber-constructed Yorkshire keel in the 1930s on the Calder and Hebble Navigation.

'Milton Princess' was the last of three boats used to transport pottery along the Caldon Canal.

kg) of coal and were carried ten at a time in narrowboats about 68 feet (20.7 m) long. After loading at a colliery the boats would travel to their destination, where a crane lifted out the loaded containers. This method speeded up the turnround of the craft and cut costs. Similar containers were also used on the Derby Canal and on certain canals in the river Severn district.

In the 1950s a modern system of containers was introduced by British Waterways in an attempt to obtain new traffic. Two sizes of container made of glass reinforced plastic were used, the large size 7 feet square (2.1 m) and 6 feet high (1.8 m) with a capacity of 4 tons, and a smaller one 5 feet square (1.5 m) and 5 feet high. The large ones were used on the north-eastern waterways for carrying goods for export and were moved from inland locations to the Humber ports in the holds of barges. The smaller ones, transported by narrowboats, operated to the Port of London from places on the Grand Union Canal. At the same time they were using aluminium open-topped box-like containers that could be carried on the back of a lorry as well as stacked several high in the hold of a barge. Excellent in use as they were, industrial troubles at the docks led to their early withdrawal, which was a waste of capital investment and a loss of traffic to the waterways. Nevertheless the waterways led the field with containerisation for the movement of goods.

Compartment boats too were introduced very early in the history of canal transport. Small rectangular floating boxes that could be fastened together into a train for movement along the waterway were used in many places, from Devon and the south-west to Wales and Shropshire. The tub boats, as the first compartment boats were called, were hauled along both by men and by horses. They provided a cheap and convenient method of transport for bulk materials over short distances.

A number of canals were built purely for tub boats, and these include Donnington Wood Canal, which had 3 tons capacity boats, and the Bude Canal in Cornwall. The latter was opened in 1819 and used primarily for carrying sea sand for use as a fertiliser. This canal, which was a unique waterway, extended for 46 miles (74 km) and had six inclined planes instead of locks. One of these, the Habbacott Incline, was 900 feet long (274 m) with a vertical lifting height of 250 feet (76 m). Boats, able to carry about 4 tons each, were fitted with iron wheels for transporting them up and down the inclined planes.

Apart from a few isolated instances, the use of tub boats had been of only moderate value, principally because of the lack of suitable power. The Aire and Calder Navigation's engineer, W. H. Bartholomew, faced with the problem of transporting coal quickly and cheaply to Goole, decided that such a system, properly adapted for their requirements, might enable them to compete with the railways. With the use of powerful steam tugs, it would be possible to use much larger compartment boats than had ever been contemplated before. To speed up the turnround and save on the lengthy and expensive process of manual unloading, he proposed using an hydraulic hoist to lift the boats, loaded with about 25 tons of coal, out of the water and tip their contents directly into the holds of ships. Operation of the new system began in 1864 and proved to be a success. Continual improvements were made and eventually the standard size of the compartment boats, which commonly became known as Tom Puddings, became 20 feet by 15 feet (6.1 m by 4.6 m), each capable of holding 40 tons. The end bulkheads of these metal compartment boats were designed so that each one was attached to its neighbour, so that when under way no side movement occurred. They were towed by tugs in trains of up to twenty at a time, and over the years many millions of tons of coal have been transported on the navigation. The hoists proved to be a success, and five were constructed, though only two now remain. After the waterways were nationalised the steam tugs were replaced by modern diesel craft.

During the 1970s the movement of coal was replaced by smoke-

Modern vessels for improved waterways: a push tug and compartment boats under way on the Aire and Calder Navigation.

less fuel and for a number of years in the region of 200,000 tonnes annually was shipped. In the early 1980s loading was concentrated at Doncaster on the Sheffield and South Yorkshire Navigation, from where the fuel was shipped via the New Junction Canal to Goole at the eastern extremity of the Aire and Calder Navigation. Finally, during the 1985-6 fiscal year only a little over 37,000 tonnes was carried in the Tom Puddings and they were withdrawn from service. Since then adapted boats have been used to test the feasibility of transporting 20 feet (6.09 m) long ISO containers in trains of fourteen, towed inland from Goole. Despite the fact that the idea for adapting Tom Puddings to carry containers was good, support for the scheme was not forthcoming. The result is that apart from the retention of a few Tom Puddings as museum exhibits, the rest have been scrapped.

On the Aire and Calder Navigation in 1967 a revolutionary new system of inland water transport was inaugurated, using new compartment boats 56 feet long (17.1 m) and 17 feet 3 inches wide (5.25 m). These, with a carrying capacity of 165 tons each, were built to transport coal to a new power station. New diesel-powered 'push' tugs were also introduced to move the compartment boats, usually three at a time, between colliery loading staithes and the power station. A large hoist capable of discharging a boat in less than nine minutes was built at the power station, and every year since then over one million tons of coal have been transported by this method, developed and operated by a private company.

In September 1970 the British Waterways Board launched its first 'push' tug, *Freight Pioneer*, and the second one *Freight Trader*, followed shortly afterwards. These tugs were used with nine new compartment boats, each 55 feet long (16.8 m) and capable of transporting 145 tons of general cargo between the ports and Rotherham. A company, Bacat (GB) Limited, aware of the potential of compartment boats, built sixty-three of them of the same size as those in use with the Board. These were loaded inland and moved to the estuary, where they were taken on board a special ship for movement into Europe. This dispensed with the need for unnecessary transhipment at the ports. This development, which promised so much for the inland waterways in the north-east, again met with the disapproval of certain trade unions. As a result both the company and the Board's commercial activities were blacked for many months. The private company could not take this forced inactivity indefinitely and closed down, so the operation that showed so much promise and would have brought millions of tons of new traffic was lost.

The British Waterways Board Freight Services Division, which had purchased additional tugs to move the Bacat compartment boats along the inland waterways, afterwards bought twenty-three of them. These, added to the Board's own fleet of nine, formed the backbone of their carrying activities in the area. This state of affairs continued until April 1988 when the Freight Services Division was disbanded and their carrying fleet disposed of. Twenty-three compartment boats, one conventional motor barge *Maureen Eva* and a number of tugs started a new lease of life joining the fleets of private carriers. The withdrawal of the Board's freight activities along the inland waterways should enable them to concentrate on servicing the network upon which the private vessels operate and benefit inland waterway transportation.

The Anderton Lift seen from the river Weaver.

7. Places to visit

National Grid references are given for those place names which might otherwise be difficult to locate.

Anderton Lift, Cheshire

Unique on Britain's waterways, the Anderton Lift is the connecting link between the Trent and Mersey Canal and the river Weaver, 50 feet (15 m) below. A gigantic structure, it was opened in 1875. The lift, now a scheduled Ancient Monument, has been out of operation since autumn 1983. Restoration work is currently underway to restore this vital navigational link.

The lift was built on an island in the river and an aqueduct, 162 feet 6 inches long (49.53 m), connected the top of the lift with the Trent and Mersey Canal. The lift, with two tanks or caissons supported on hydraulic rams 3 feet (0.9 m) in diameter, began work in 1875. One tank was always up when the other was down, and, for them to operate, gates at the ends of the tanks were shut and the lift was set in motion by pumping a small amount of water from the lower tank. The descending tank, capable of holding a pair of narrowboats, moved into the new position at the bottom.

After thirty years of service, extensive repairs were required and these gave the opportunity to redesign the working parts. An electric motor was installed and the tanks were suspended by means of wire ropes from an overhead arrangement of cogs, pulleys and counterweights. The alterations, completed in May 1907, allowed each tank to be operated independently, providing a more satisfactory way of working. Each tank is 75 feet long (22.9 m) and 15 feet 6 inches wide (4.7 m), holds a 5 feet (1.5 m) depth of water and has an inclusive weight of 250 tons.

Avon Lock, Tewkesbury, Gloucestershire

On a pleasant summer afternoon there is no better place for a waterway enthusiast to spend an hour or so than on the riverfront of the town of Tewkesbury. Particularly interesting is the vicinity of the Avon Lock, which is the entrance to the Lower Avon from the river Severn. The lock is now electrically operated and this assists in the swift movement of the numerous pleasure craft that often queue up to pass through.

Above the lock many and various types of pleasure craft will be seen moored, while below, on the channel from the Severn, may be seen large modern commercial barges owned by the nearby flour mill.

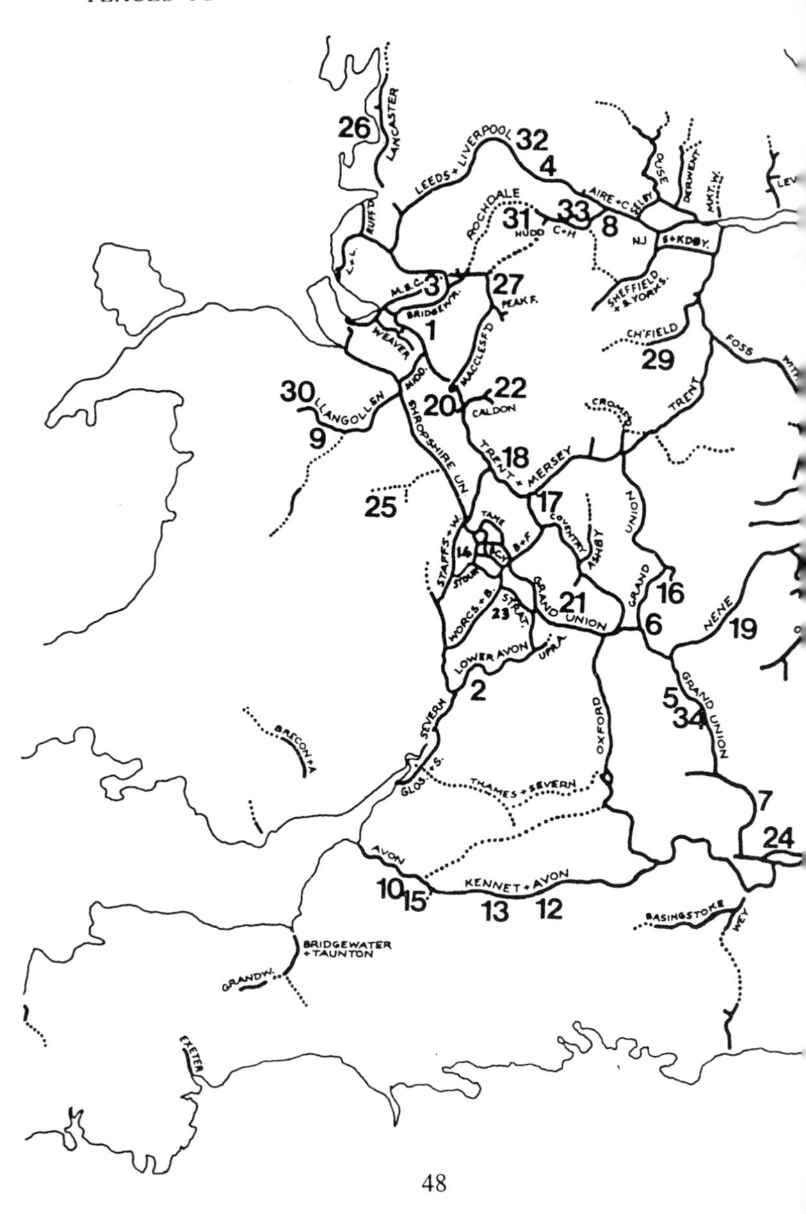

LANCASTER
26
LEEDS + LIVERPOOL
32
4
RUFFD
L+L
ROCHDALE
31
HUDD
C+H
33
AIRE + C
SELBY
8
OUSE
DERWENT
MKT. W.
NJ
S+KDBY.
SHEFFIELD + S. YORKS.
M.S.C.
3
27
PEAK F.
BRIDGEWTR.
1
WEAVER
CH'FIELD
29
FOSS
MIDD.
MACCLESFD
22
30
LLANGOLLEN
20
CALDON
9
SHROPSHIRE UN.
TRENT + MERSEY
18
CROMFD
TRENT
25
17
COVENTRY
ASHBY
UNION
TAME
STAFFS + W.
14
B+F
STOUR
GRAND UNION
21
GRAND
16
WORCS + B.
STRAT.
23
6
NENE
19
LOWER AVON
UPR A.
2
GRAND UNION
5
34
SEVERN
OXFORD
BRECON + A
GLOS + S.
THAMES + SEVERN
7
24
AVON
10
15
KENNET + AVON
13
12
BASINGSTOKE
WEY
BRIDGEWATER + TAUNTON
GRAND W.
EXETER

Canals and river navigations of England and Wales. Dotted lines denote disused waterways. The numbers indicate the locations of the places to visit described in Chapter 7: 1 Anderton Lift; 2 Avon Lock; 3 Barton Swing Aqueduct; 4 Bingley Five Rise; 5 Blisworth Tunnel; 6 Braunston; 7 Cassiobury Park; 8 Castleford Junction; 9 Chirk Aqueduct and Tunnel; 10 Claverton Pumphouse; 11 Crinan (see map on page 50); 12 Crofton Pumping Station; 13 Devizes Locks; 14 Dudley Tunnel; 15 Dundas Aqueduct; 16 Foxton Locks; 17 Fradley Junction; 18 Great Haywood Junction; 19 Guillotine gates on river Nene; 20 Harecastle Tunnel; 21 Hatton Flight; 22 Hazelhurst Junction; 23 Kings Norton Junction; 24 Little Venice; 25 Longdon Aqueduct; 26 Lune Aqueduct; 27 Marple Aqueduct and Locks; 28 Neptune's Staircase (see map on page 50); 29 Osberton Lock; 30 Pontcysyllte Aqueduct; 31 Sowerby Bridge Basin; 32 Springs Branch; 33 Stanley Ferry Aqueduct; 34 Stoke Bruerne.

Places to visit on Scottish waterways: 11 Crinan; 28 Neptune's Staircase.

Barton Swing Aqueduct, Greater Manchester

This aqueduct conveys the Bridgewater Canal over the Manchester Ship Canal, and it is one of the great engineering marvels of the late Victorian period. On the same site as the present aqueduct was the famous Brindley three-arched stone aqueduct. The construction of the Ship Canal made the old aqueduct obsolete. It had given service for 130 years, but it had to be replaced to allow large ships to pass. Edward Leader Williams, of the Ship Canal company, designed the present structure, a swing aqueduct. It includes a tank of wrought iron some 235 feet long (71.6 m) that provides a navigable width of 18 feet (5.5 m) and a depth of 7 feet (2.1 m) of water for the Bridgewater Canal. The tank holds a total of 800 tons of water and is positioned on sixty-four cast iron rollers that act as a central pivot. The entire moving structure weighs 1,450 tons.

Because the aqueduct swings with a full tank of water, the ends must be made leakproof before it is moved. This is done with gates similar to lock gates, which open back either to the tank or the canalside when not required, for the canal must be sealed as well. In between the gates of the tank and the canal is a small gap that

allows the tank to swing. When the aqueduct is in use this gap is bridged with a watertight seal, which is a 12 ton rubber-shod iron wedge of the same section as the canal and is moved into place by hydraulic rams. The aqueduct, completed in 1894, swings on the central pivot and roller path, until it lies at right angles to the Ship Canal to allow ships with high masts and superstructure to pass.
Grid reference: SJ 767976.

Bingley Five Rise, West Yorkshire

On the Leeds and Liverpool Canal in Bingley is the famous Bingley Five Rise. It takes about half an hour to lift a vessel the 60 feet (18.3 m) to the top. After the strenuous effort of working up the famous flight there is the pleasure of a 16 mile (26 km) lock-free pound to Gargrave.

Designed by John Longbottom, the locks are a very impressive sight. One of the wonders of the waterways, the Bingley Five Rise attracts a vast number of visitors throughout the year. They range from groups of schoolchildren during term time to large crowds of family groups on summer bank holiday weekends. So novices should be warned to practise lock operation and rope throwing beforehand, especially the latter. It is surprising how entertaining it can be to watch someone on a boat at the bottom of a lock trying for the first time to throw a line ten to twelve feet straight up to an exasperated crew member above.
Grid reference: SE 108399.

Blisworth Tunnel, Northamptonshire

Along the towpath from Stoke Bruerne is the Blisworth Tunnel, wide enough to allow two boats to pass. It is nearly 1.75 miles long (2.8 km) and, after several attempts, was completed in 1805. Because it was found so difficult to dig, with continuous flooding and roof falls, the first attempt was abandoned in 1796. In 1802 a different route was started, with adequate provision for drainage for the safety of the construction teams, and so success was achieved. Meanwhile the canal to both ends of the tunnel had been completed and boats were operating. The gap in the system was bridged and the movement of cargoes made possible by the construction of a tramroad over the hillside, which was used until the completion of the tunnel.

The tunnel does not have a towpath and as a result the horses had to be led over the top to meet their boats at the far side. For many years the boats were legged through, a method whereby two men lay on a cross plank and walked their way through, pushing with their feet against the walls. On the hillside over the top of the tunnel a number of ventilator shafts can be seen. These enable one to trace the route of the subterranean waterway. Originally nineteen shafts were sunk for constructional purposes, but these were filled in when the tunnel was completed. Later a number of the shafts were reopened to provide ventilation for the boat crews, because in 1861, after the introduction of steam-powered boats, two men were suffocated. Following structural problems in 1980 a major rebuilding programme replaced the centre section of the tunnel with preformed concrete sections.

Braunston, Northamptonshire

Braunston is a well-known canal centre, with a flight of six locks that are the scene of much activity during the summer months. A little beyond the top lock is the Braunston Tunnel, which bores through the Northamptonshire Heights. Because of quicksand it was an extremely difficult tunnel to dig, but success was achieved in 1796. It is 2,042 yards long (1,867 m) and of a wide section, as are all the tunnels on the Grand Union Canal route to London.

The village is centred around the canal and for many years depended upon it for employment, for it was a very busy place, with boatbuilding and repairs undertaken at several yards. The association with boatbuilding, services and repairs still continues but is now mostly confined to the pleasure craft trade, with extensive moorings, slipways and suchlike. There is a triangular junction at the northern end of the village, the connection with the Oxford Canal. In the centre of the junction is a small island reached by a pair of elegant matching cast iron bridges that carry the towpath.
Grid reference: SP 539658.

Cassiobury Park, Watford, Hertfordshire

The Grand Union Canal goes through Cassiobury Park, seat in the seventeenth century of the Earl of Essex. It is now a public park of nearly 900 acres (364 ha), which includes Whippendell Woods, and is complete with playing fields and a golf course. The canal is not the only watercourse in the park for the river Gade also flows through it and alongside the canal. The canal adds to the beauty of the area, as the builders intended, for it winds like a river and the many mature trees, including limes, complete the scene. The canal runs through the park for almost 2 miles (3 km), with four locks and a number of bridges in the park. The bridges are of good design, but number 164 is exceptional. Well built of honey-coloured stone, it has one central arch over the canal, with two small ones on either side for pedestrians. Along the top of the bridge is a well proportioned balustrade of the same stone.

Castleford Junction, West Yorkshire

If you are fascinated by the sight of commercial craft under way, then a visit to Castleford Junction is a must. It is situated a little to the north of the town, where the rivers Aire and Calder meet at a four-way junction. From there the combined rivers continue their journey to the sea as the river Aire, flowing in a wide loop, on which there is a weir. The main route of the Aire and Calder Navigation bypasses the weir by an artificial cut, which is about half a mile (800 m) in length. It begins at the junction with an electrified flood lock, which is most unusual for it has a curved chamber. To steer a large modern vessel through the lock without touching the sides requires great skill. Operation of the lock and control of the traffic flow is maintained by a lock keeper, who is positioned in an overhead control cabin.

Near the lock are the Navigation's old toll office and the modern offices of the Area Manager while on the other side of the canal are a utility-built warehouse and pleasure craft facilities. These are located alongside a layby, where overnight stops are regularly made by both commercial and pleasure craft. A little way along the canal section is a road bridge, which carried the A656 road from the town. On the other side of the bridge is a repair yard, where both push tugs and compartment boats, along with conventional barges, are to be seen either awaiting or undergoing repairs. At the far end of the canal section is the electrified Bulholme Lock, the largest on the navigation. It has a chamber 460 feet long (140 m) and is fitted with intermediate gates, so that it can accommodate just one vessel at a time if need be.

Chirk Aqueduct and Tunnel, Clwyd

It is suggested that the towpath walker joins the towpath of the Llangollen Canal near Morton bridge, number 17, where the A5 road passes over the waterway. The canal runs through the beautiful Ceiriog valley and the wooded hillsides along the way make it a most pleasant walk. After about a mile or so the canal makes a right-angled turn, and ahead is the Chirk Aqueduct. It has a trough 9 feet wide (2.7 m), made up of flanged cast iron plates bolted together. The trough, encased in mellowed warm-looking stone, is supported on arched columns of the same stone. The aqueduct was built to carry the canal over the river Ceiriog some 70 feet (21 m) below. Alongside the canal, crossing the river, is a railway viaduct that carries the line from Shrewsbury to Chester. Unfortunately as it is at a higher level than the aqueduct, the viaduct, built forty years after the canal, overshadows the grandeur of the aqueduct.

Immediately beyond the aqueduct the canal widens out to provide a passing point for vessels before it enters Chirk Tunnel. The

459 yard (420 m) tunnel was completed in 1802. It is of narrow section, brick-lined throughout, and is provided with a towpath, complete with a handrail to separate it from the canal. This provides the non-boating enthusiast with the opportunity to walk through and experience the sensation of a tunnel, although as it is not very long daylight can be seen at both ends throughout its length.

Claverton Pumphouse, Avon

To supply water to the Kennet and Avon Canal, a pump, housed in a fine-looking building of Bath stone, was built at Claverton by John Rennie. It commenced work in 1813, lifting 100,000 gallons (450,000 litres) per hour up to the canal from the river Avon, which is over 50 feet (15.2 m) below. Initially using a waterwheel 25 feet wide (7.6 m) and over 19 feet in diameter (5.8 m), it was later modified because of continuous problems with the bearings. Declining traffic on the canal led to the disuse of the pump, and it was only after considerable restoration work in the 1970s by volunteers that the massive waterwheel and pump were brought back into use.

The pumphouse is only 3 miles (4.8 km) from Bath, and between March and October a number of 'pumping' weekends are staged for the benefit of visitors. It is also open for inspection on Sundays during the same period.

Crinan, Strathclyde

The village of Crinan, after which the short Scottish seaway, the Crinan Canal, is named, is located on the western coast of the Kintyre Peninsula, in one of the finest situations in the British Isles. It can be best appreciated when viewed from the top of a nearby high cliff. From there the entrance lock, two canal basins and the surrounding small buildings can be seen to advantage. There is a small lighthouse at the entrance to the Crinan Canal sea lock, and above the locks are the two small basins that provide safe moorings for sailing craft. The small village consists of a number of cottages, one shop and a hotel clustered together on one side of the waterway. The whole, especially early on a summer's evening, combines to make one of the most picturesque views possible.

Crofton Pumping Station, Wiltshire

This station is situated alongside the Kennet and Avon Canal between Burbage and Great Bedwyn. It is about 5 miles (8 km) from the A4, and to get there leave the A4 at Froxfield near Hungerford and follow the signposts for Great Bedwyn. Turn right there and the pumping station is two miles beyond the village on the road to Burbage. The Crofton Pumping Station houses two

Crofton Pumping Station on the Kennet and Avon Canal.

early nineteenth-century beam engines, one of which is the oldest working beam engine in the world, built in 1812 by Boulton and Watt. The other was built in 1814 by Harveys of Hayle. They can be viewed every Sunday from April to October but are only in steam on certain dates.
Grid reference: SU 264626.

Devizes Locks, Wiltshire

The Caen Hill flight at Devizes on the Kennet and Avon Canal with twenty-nine locks is the second longest flight in Britain. The locks were derelict for a number of years but restoration of the canal is now completed. The locks numbered 22 to 50 lift the canal up the Caen Hill and extend over a distance of 2 miles (3.2 km), but sixteen of them are positioned very close together.

The former bonded warehouses on Devizes Wharf have recently

The canal wharf at the Black Country Museum, Dudley.

been reconstructed by the Kennet District Council and renamed the Devizes Wharf Canal Centre. They are now the headquarters of the Canal Trust. Also included in the centre is the Trust shop, together with a tourist information centre, a museum, and an interpretation centre for the countryside.
Grid reference: SU 985615

Dudley Tunnel, West Midlands

The Dudley Canal Tunnel was completed in 1792 and for almost 150 years was a hive of activity. During the 1950s traffic through the tunnel gradually declined, until in 1962 the entrances were boarded up. Concern over its loss resulted in the formation of the Dudley Canal Trust Preservation Society. With the cooperation and assistance of the British Waterways Board and local councils and after much physical effort they succeeded in reopening the tunnel in 1973. Because of ventilation problems both petrol and diesel engines are banned. As a result most people travel on board the battery-operated boat of the Dudley Canal Trust when they wish to explore the fascinating old tunnel. Its length is officially given as 3,172 yards (2,900 m) but a recent check on this by Trust members produced the figure of 3,154 yards (2,884 m). Some people doubt whether it can be called one tunnel, for it is split into three separate tunnel sections, with two small intermediate basins which are open to the sky. A visit to the Black Country Museum, adjacent to the tunnel, is well worthwhile.
Grid reference: SO 942910.

Dundas Aqueduct, Avon

The Dundas Aqueduct, a little to the east of the Claverton Pumphouse, is one of the finest aqueducts ever built. Like the Avoncliffe Aqueduct not far away, it carries the Kennet and Avon Canal over the upper reaches of the Bristol Avon. The aqueduct, designed by Rennie, combines the best of classical eighteenth-century architecture with the Georgian style of Bath. A magnificent structure, it can be seen to best advantage from the hillside overlooking the valley in which it stands.
Grid reference: ST 784625

Foxton Locks and Inclined Plane, Leicestershire

In July 1896 the Grand Junction Canal Company decided to build an inclined plane at Foxton, to provide an efficient and quick method of transporting boats up and down the steep hillside there. The inclined plane, machinery and boiler house cost £37,500 and were opened on 10th July 1900. Two tanks or caissons, each 80 feet (24 m) by 15 feet (4.6 m) and capable of holding a pair of narrowboats, travelled sideways on rails up and down the inclined plane. One would ascend as the other descended, counterbalancing each other on the twelve minute journey.

Operation of the inclined plane was a success, resulting in a saving of time and labour, in comparison with the forty-five minutes needed to negotiate the locks. These were then allowed to fall into disrepair. The lift was only operational during the day, so in 1908 it was decided to repair the locks to allow overnight traffic to proceed. It was considered uneconomical to keep the lift staffed and with a full head of steam for the small overnight traffic. In October 1910 the company decided to close the lift down. High operating costs and a need to modify the rail system on the incline prompted this action. The lift was last used in March 1911, but it was not until 1927 that it was dismantled and sold for scrap.

The Foxton Inclined Society, formed in the 1970s, has reconstructed the engine house and created a permanent exhibition in a new visitors centre. Meanwhile work on reconstructing the historic inclined plane continues.

Access to the canal for the towpath walker to view the site and the Foxton Locks is from bridge number 60 on the main line above the locks. Near this access point is a convenient car park and picnic area. The Foxton Locks are narrow-beamed ones, arranged in two separate staircases of five chambers each, with a central passing point. Instead of passing directly from one lock to the next, the water passes into side ponds. At each end of a lock chamber there is just one ground paddle gear. They are painted different colours. The red one controls water flow from the side

The southern end of the Harecastle tunnel on the Trent and Mersey Canal. The light in the tunnel is a boat headlight. The building over the tunnel entrance houses an extractor fan used to dispel the diesel fumes.

pond into the chamber, whilst a white one empties the lock chamber, allowing the water to flow into the side pond, to be used in the lock below.
Grid reference: SP 692896.

Fradley Junction, Staffordshire

Fradley Junction is usually a very busy place as it is about halfway up a short flight of five locks, which lift the Trent and Mersey Canal over 35 feet (10.7 m). Also it is the place where two important waterways, the Coventry Canal and the Trent and Mersey Canal, meet. Opposite the junction is the Georgian Swan Inn, one of the best known canalside pubs. Nearby is some former stabling, now used as a canal shop and base for Swan Line Cruisers. Everything at the junction has to do with the canal, for the village of Fradley itself is some distance away. Nearby are the local British Waterways Maintenance Section repair yard and moorings for many pleasure craft. Away from the hustle and bustle of the main roads the junction can be a very peaceful spot indeed.
Grid reference: SK 141140.

Great Haywood Junction, Staffordshire

Great Haywood is an eighteenth-century canal junction, where the Staffordshire and Worcester Canal joins the Trent and Mersey Canal. At the junction there is a slender mellow brick towpath bridge, which is unusually wide and has a slightly pointed arch. This crosses over the most northern point of the Staffordshire and Worcester Canal. Alongside is Great Haywood Wharf, used since 1972 as a hire cruiser base by Anglo Welsh Narrowboats. It is an interesting area, well worth exploring, with many waterway features. Only a short distance along the Staffordshire and Worcester is a small brick-built lengthman's cabin, with arched windows that have iron latticework window frames, and nearby is a small aqueduct carrying the canal over the river Trent. Less than a mile further along the canal is Tixall Wide, a delightful stretch of water where the canal broadens out to look exactly like a lake, and where one can see many different species of birds. Across the water to the north is a splendid Elizabethan gatehouse of the former Tixall Hall.

A short walk in a southerly direction from the junction along the Trent and Mersey Canal is Haywood Lock, and a little beyond the lock is a pretty and popular mooring spot for vessels. This is near Shugborough Hall, which is over to the right. Access to it is across a narrow former packhorse bridge.

Guillotine gates on the river Nene, Northamptonshire

The river Nene was made navigable inland to Northampton in August 1761. In the 1930s the Nene Catchment Board was established and took over responsibility for navigation, though primarily concerned with drainage. A major scheme for improvement was put in hand by the Board. It included the reconstruction of every lock on the river, where they installed a guillotine gate at the bottom end of each lock. Originally thirty seven locks on the river were fitted with these gates. They each required over 150 turns of the windlass to lift the gate high enough for a boat to pass under, but recently some have been electrified.

Harecastle Tunnel, Staffordshire

Access to the northern end of the Harecastle Tunnel, on the Trent and Mersey Canal, can be gained via the towpath from Kidsgrove. There are two tunnel entrances to be seen but the one on the right, now closed, was the first to be built. It was engineered by James Brindley. Started in 1766, it took over eleven years to complete. As there was no towpath, all boats had to be legged through. Subsidence was a continual problem, reducing the headroom considerably in many places.

The passenger boat 'Jenny Wren' at Little Venice in London.

In 1822 Thomas Telford advised the Trent and Mersey directors to build a second tunnel, parallel to the first. As a result, two years later he commenced work on the second one. By that time tunnel construction techniques had improved and the second tunnel, with a towpath, was opened after only three years work. The original tunnel, despite its problems, remained in use until 1918. During the 1950s subsidence was affecting the second tunnel so much that a series of repairs had to be undertaken and sections of the towpath were removed. In 1973 the tunnel was closed to traffic and remained so until early in 1977. In that time extensive roof repairs were carried out to sections of the 2,926 yard (2,676 m) tunnel.
Grid reference: SJ 844528.

Hatton Flight, Warwickshire

The Hatton flight of twenty-one locks extends over a distance of 2 miles (3.2 km). They lift the Grand Union Canal to a height of 337 feet (103 m) above sea level. Before the 1930s the locks were only passable by narrow-beamed craft but a modernisation scheme, designed to make the waterway a wide-beamed one all the way from London to Birmingham, resulted in the construction of new locks at Hatton. A few chambers of the former narrow locks can still be seen. The locks now in use have large gates and low geared paddle gearing, and their operation can be hard work for a small crew.
Grid reference: SP 241669 (top lock).

Hazelhurst Junction, Staffordshire

The present junction and most of the canal in its vicinity are now in a different position from when they were first constructed, because of the Leek railway built in 1841. The Caldon Canal, located in beautiful hilly countryside, has two terminals and Hazelhurst Junction is where the two branches part. From the junction, the canal to Froghall passes under an interesting cast iron footbridge and alongside a lock keeper's cottage. Then it drops 24 feet (7.3 m) through three locks and also passes underneath an aqueduct. Just after the last lock there is a feeder and from there the route of the canal reverts back to the original line of the canal. The canal section to Leek, after parting company with the Froghall section, continues at the same level and parallel to it for a few hundred yards. Then, turning sharply to the left, it passes under a stone bridge to go over the other canal by the aqueduct. This Hazelhurst Aqueduct is a well proportioned brick-built structure and is one of the most majestic ever built of burnt clay.
Grid reference: SJ 954537.

Kings Norton Junction, West Midlands

Situated in the suburbs of Birmingham and surrounded by urban sprawl, Kings Norton is basically still a village and even retains a village green. Here in a peaceful setting the Northern Stratford-on-Avon Canal connects at a T junction with the Worcester and Birmingham Canal. From the Kings Norton Junction it is only a few yards along the Stratford-on-Avon Canal to the Kings Norton Stop Lock, which is no longer used, although it is still complete with its old guillotine gate, its framework and its ancient mechanism.
Grid reference: SP 053793.

Little Venice, London

In the heart of London is Little Venice, where the Regent's Canal and the Paddington Arm meet to form Browning's Pool, named after Robert Browning. It is a very attractive area with many Regency houses, several interesting canal cottages and a former toll house.

Some time ago the walls along Bloomfield Road and Maida Avenue were replaced with railings. This revealed the full beauty of the tree-lined section of canal up to Maida Hill Tunnel and became the first stage of a canalside walk. Later the process was repeated along Delamere Terrace, which became one of the early attempts to open up the urban towpaths in London.

A waterbus service from Little Venice to the Zoo has operated every summer from the late 1950s.

Longdon Aqueduct, Shropshire

The Shrewsbury Canal, opened in 1797, was primarily built for the working of trains of tub boats and so it had locks of the unusual size of 6 feet 7 inches wide (2.0 m) and 81 feet long (24.7 m). On its route the canal had to cross over the river Tern, where the engineer Josiah Clowes, had decided to build an aqueduct of brick and stone. But Clowes died as the abutments of the structure were competed. Telford was then commissioned to complete the project and did so with William Reynolds. The aqueduct was opened in March 1795 and fitted with a cast iron trough supported on cantilevered iron legs. This aqueduct is still standing, demonstrating that lightweight iron structures could be prefabricated and erected more easily than a masonry structure. This was only the second aqueduct with a cast iron trough to be built, the first one being Benjamin Outram's single-span Holmes aqueduct on the Derby Canal, opened one month earlier. One drawback to the Longdon Aqueduct was that it was made only a few inches wider than the boats that used it. So there was little room for displacement of the water and a great deal of effort was needed to move a boat across the aqueduct.

Lune Aqueduct, Lancashire

The most magnificent structure on the Lancaster Canal is the

Marple Aqueduct on the Peak Forest Canal.

Lune Aqueduct, which spans the river Lune on the outskirts of Lancaster. It took four years to build and was completed in November 1797. Its construction was not without problems, for a considerable number of coffer dams were required, with steam pumps used to keep them clear of water. The aqueduct is made up of five identical arches, each one with a 70 feet (21 m) span, giving the structure a total length of 664 feet (202 m). The piers between the arches rest on piles driven 30 feet (9 m) into the river bed.

One of Rennie's earlier designs, the aqueduct is notable for its attractive proportions and the ornate decoration. The balustrades are of local stone and below the elegant cornice on each side there is an inscription. The total cost of the aqueduct was nearly £50,000, enough to dissuade the Lancaster Canal Company from erecting their proposed Ribble aqueduct.
Grid reference: SO 483639.

Marple Aqueduct and Locks, Greater Manchester

Marple is on one of the highest lengths of navigable canal in Britain, where the Peak Forest Canal and the Macclesfield Canal meet. At the junction, over the Macclesfield Canal is a wonderful example of a stone turn-over bridge, while on the Peak Forest Canal, alongside the junction, is the top lock of the sixteen-lock Marple flight, opened in 1804. The locks are well built with stone chambers and many little access bridges at the bottom ends. One even has its own little towpath tunnels, one for the horse and one for the boat crew.

At the bottom of the flight is the very impressive Marple Aqueduct, which crosses over part of the Etherow valley, where it is particularly deep and well wooded. A railway viaduct built alongside rather diminishes the impact of the awe-inspiring structure, although this will be forgotten after a walk down the riverbank to the valley bottom. In the arch abutments are round apertures, made to reduce the overall weight but also to improve the appearance. The aqueduct is a three-arched stone structure. There is a stone parapet on the towpath side but not on the canal side, where there is a drop of 100 feet (30 m) to the river below.

Neptune's Staircase, Banavie, Highland

In Scotland, near the western end of the Caledonian Canal, is the spectacular staircase of eight locks at Banavie, which are more commonly known as the Neptune's Staircase. They are an impressive sight and of enormous proportions, providing a change of level of 64 feet (19.5 m). Designed by Telford, they can accommodated sea-going craft 150 feet by 30 feet (45.7 m by 9.1 m). All the locks were mechanised in the late 1960s.

Osberton Lock, Nottinghamshire

Osberton Lock is a little haven of peace on the Chesterfield Canal. Here are no gigantic or majestic structures, only a narrow lock and a small brick-built lock keeper's cottage alongside the canal. It is a canalside paradise, like many hundreds of other spots throughout the canal system, but it is a firm favourite of mine. To get there by road, proceed along the A1 and take the A620 road for Worksop. After about 1.5 miles (2.4 km), turn right at the signpost for Scofton, and another couple of hundred yards will bring you straight to the lock.

At the lock, first stand and stare and take in the scene, and then take a stroll along the towpath upstream. As you walk along you will surely appreciate the beauty of nature, especially if it is late spring, for all around you and beneath the trees will be a carpet of bluebells. Because cuckoos are so often heard on this towpath, the lock for many people is known as Cuckoo Lock. Afterwards return to the lock, and then walk along the bridlepath to Scofton. There are no pubs or shops, but make a point of exploring the church and its surroundings and in the churchyard stand and look at the fine view over the green meadows towards Osberton Hall.
Grid reference: SK 631801.

Pontcysyllte Aqueduct, Clwyd

Located near Trevor in North Wales, the Pontcysyllte Aqueduct carries the Llangollen Canal over the river Dee. The aqueduct is made up of a trough of cast iron flanged plates 1,007 feet in length (306.9 m). This rests on numerous slender stone columns 121 feet (36.9 m) high above the river. The towpath is suspended over the trough and has railings on the outside edge, while the canal edge is without any barriers whatsoever, and so from the outside edge of a boat it is a sheer drop to the river and valley below. The aqueduct is an amazing structure, especially when viewed from near the old three-arched road bridge upstream on the Dee.
Grid reference: SJ 271420.

Sowerby Bridge Basin, West Yorkshire

Some of the best stone canalside warehouses ever built were constructed alongside the Calder and Hebble Navigation terminal basin at Sowerby Bridge. The first warehouse was erected in the 1780s and a continual increase in trading resulted in the building of further warehouses, forming the present scene, with warehouses numbered 1 to 4, surrounded with granite sets, along with office and stabling accommodation. They provided full facilities for warehousing and distribution, totalling 3,673 square yards (3,071 sq m) of covered accommodation and 1,800 square yards (1,505 sq

Stoke Bruerne on the Grand Union Canal.

m) of outside storage. The last barge delivery to the warehouses was in September 1955. Afterwards the warehouses were left empty and derelict, but in the 1970s life returned, and now the basin is used for hire boats and moorings. Following the ongoing restoration of the Rochdale Canal, Sowerby Bridge now has new lock links to the largest of the cross Pennine navigations.

Springs Branch, Skipton, North Yorkshire

The canal at Skipton is worth exploring, especially the Springs Branch. It is a little over half a mile (800 m) in length and branches off the main line of the Leeds and Liverpool Canal within the town area. After the junction the canal runs at the bottom of a ravine or cutting 100 feet (30 m) below but alongside Skipton Castle. The Springs Branch section of canal was built to carry limestone, which was loaded into vessels at a chute at the very end of the canal. This material was transported to the canal from the quarries by rail. The last boats were loaded in the early 1950s and since then most of the short waterway has remained a backwater. The towpath walker, after the end of the canal, may like to continue the walk, and this can be done by going through Skipton Woods and afterwards visiting the castle before returning to the town centre.

Stanley Ferry Aqueduct, West Yorkshire

The aqueduct is located on the Wakefield section of the Aire and Calder Navigation and carries the navigation over the river Calder. The aqueduct, opened on 8th August 1839, was designed by William Leather, at a cost of nearly £50,000. It consists of a trough 180 feet in length (55 m), which holds 940 tons of water. At each side of the 24 feet wide (7.3 m) trough are two bow-shaped arcs, with suspended rods that hold the trough. As the first stage of upgrading the Wakefield section of the navigation to a 700 ton standard, work commenced on a new concrete structure late in 1979. The new aqueduct, alongside the old, makes a deeper draught available and so allows bigger vessels to operate. The old aqueduct, a marvel of engineering, is now scheduled as an ancient monument and will remain where it stands.
Grid reference: SE 355231.

Stoke Bruerne, Northamptonshire

Stoke Bruerne is a famous canal centre, which all waterways enthusiasts should visit. The centre of attraction is the Canal Museum, opened in 1963. It is housed in a stone building which was formerly a corn mill. On the canalside near the museum are canal shops, a genuine old boatman's pub and a number of cafes. Alongside the museum is the top lock of the Stoke Bruerne flight of seven, which is most interesting, having two separate chambers, one of which is now used as a dry dock and houses a former Glamorgan Canal weighing machine. Stoke Bruerne is on the Grand Union Canal, which is very busy throughout the summer months.

8. Popular waterways

Aire and Calder Navigation

Length: Goole to Castleford 24 miles (39 km); Castleford to Wakefield 7.5 miles (12 km); Castleford to Leeds 10 miles (16 km).

Locks: Goole — Castleford three, plus two flood locks; Castleford —Wakefield three, plus one flood lock; Castleford—Leeds six, plus one flood lock.

The first Act for the navigation was in 1699, with the first boats arriving in Leeds during 1700 and in Wakefield in 1701. It is a busy commercial waterway, on which modern large capacity craft, including trains of compartment boats propelled by 'push tugs' and also bulk fuel craft that can carry up to 700 tons, are regularly seen. Goole, on the river Ouse, is the terminal port, where a large hoist that lifted the Tom Puddings out of the water can still be seen.

The waterways museum at the Sobriety Centre relates the fascinating history of the Aire and Calder Navigation. It is also the base for a trust working to develop the last boat hoist as a visitor centre. Original compartment boats are displayed outside.

The waterway does not run through beautiful country, but it provides an interesting journey with much to see. The eastern section passes through many miles of agricultural land, while along the western section there is a mixture of urban sprawl and industrial sites. The journey along the western section is becoming, however, more pleasant, especially on the 4 mile (6.4 km) of the river Aire section below Castleford, as well as on the section of the navigation between Castleford and Wakefield. For many years a good majority of the banks consisted mostly of colliery waste, but changes have and are taking place, mostly because of environmental improvements aided by the healing hand of nature. This is resulting in the growth of grass, shrubs and trees, making a pleasant linear park bordering the waterway.

On the journey inland from Goole, the New Junction Canal connects at the 7.25 mile point (11.7 km). The Selby Canal route access lock is 10 miles (16 km) further inland at Knottingley, which is a busy waterway centre, with a shipbuilder's yard and slipways as well as many trading wharves. The Ferrybridge power station complex, on the river Aire beyond Knottingley, has daily deliveries of coal that total more than 1.5 million tonnes a year. At Castleford the routes to Wakefield and Leeds separate, and this junction is very busy at most times of the day. On the Wakefield section is the unique nineteenth-century Stanley Ferry Aqueduct, with the new concrete aqueduct alongside which now

The Royal Armouries Museum overlooks the canal at Leeds.

carries vessels over the river Calder, while on the Leeds route are a number of bulk fuel installations. In the city overlooking the river there are still several interesting old warehouses and industrial buildings.

The most successful of the Board's commercial waterways, it produces an annual profit. As it is fitted with electric locks no physical effort is required by the crews of cruisers. Towpath walkers, on the other hand, will need a great deal of determination to keep going, for despite a lot of improvements some of the river sections still have inaccessible towpaths.

Ashby Canal
Length: 22 miles (35 km).
Locks: narrow-beamed stop lock at Marston Junction.

The canal is part of a 50 mile (80 km) section of cruising waterways that is level apart from a change of only a few inches because of stop locks at Hawkesbury and Marston. They include the Ashby, the Northern Oxford to Hillmorton and part of the Coventry Canal.

The canal was constructed under an Act of 1794. Originally it extended from Marston Junction, the connecting link with the adjoining Coventry Canal, 30 miles (48 km) to Moira. 8 miles (13 km) at the top of the canal have been abandoned, leaving the navigable section to terminate a little beyond the 250 yard long (229 m) Snarestone Tunnel. At the terminal point a winding hold is provided to enable craft to turn round. A very quiet and peaceful waterway, passing through pleasant farmland, it is little used. Lock-free and without many great engineering features, the canal winds its way, generally avoiding the villages in the area. It does however pass close to Bosworth Field, the site of the battle fought in 1485 which ended the Wars of the Roses. The battlefield is accessible from the canal near Shenton.

Ashton Canal
Length: 6.25 miles (10 km).
Locks: eighteen (narrow-beamed).

The canal is an integral part of the Cheshire Ring of cruising waterways, a circular route of six different connecting waterways. It is now a busy cruising waterway during the summer months. At Manchester it connects with the 1.25 mile long (2 km) open section of the Rochdale Canal, and at Dukinfield Junction at the other end, at Ashton under Lyne, it connects with the Peak Forest Canal.

The canal received its first Act of Parliament in 1792 and was opened in 1799. It had connections with the now closed Huddersfield Narrow Canal and the Stockport Canal and remained a busy waterway until well into the twentieth century. The Ashton Canal itself became derelict in the early 1960s and would have remained closed to traffic but for the efforts of the Peak Forest Canal Society, who spearheaded restoration work that resulted in its reopening in 1974.

The canal runs through a predominantly industrial area and it offers many opportunities to study the numerous Victorian commercial buildings along its banks, some of which have features of architectural merit. Increasing problems with vandalism mean that boat movement is supervised by British Waterways staff. Details should be sought from the local office.

River Avon (Warwickshire) Navigation, Lower section
Authority: Lower Avon Navigation Trust.
Length: 24 miles 5 furlongs (40 km), Tewkesbury to Evesham.
Locks: eight, 70 feet by 13 feet 6 inches (21 m by 4.1 m); draught 3 feet 6 inches (1.07 m).

The River Avon Navigation was one of the earliest and was made navigable in the 1630s. The section of the river known as the Lower Avon became unnavigable upstream of Pershore during the Second World War. In 1950 it was purchased by Mr C. D.

Barwell for £1,500. He instigated the formation of a charitable trust to restore the river as an open navigation. Swift action followed, resulting in the restoration of Strensham Lock in the following year. Then as a result of great efforts by many people the river was made navigable to Offenham, a little upstream of Evesham. Complete restoration was officially recognised when the Lower Avon was reopened to vessels on 7th June 1964.

The river from Tewkesbury is a beautiful waterway to cruise along, with many villages and the delightful towns of Tewkesbury, Pershore and Evesham to add to the pleasures of exploration. There is much attractive scenery and there are many buildings with charm and appeal, such as Fladbury Mills and the bridges at Eckington, Fladbury and Pershore. In addition there are some fascinating locks, some of them unique structures. At particularly busy points some of the locks have been electrified and are operated by Trust staff or by members during working hours.

The river now forms part of the Avon Ring, a circular route of cruising waterways. This is a popular route and so the waterway is very busy throughout the summer months. A cruising licence is required for the waterway and this can be obtained beforehand or upon request at the manned locks at each end of the river. A licence is also available that allows unlimited cruising for a specific period on the two connected waterways that do not come under the control of British Waterways.

River Avon (Warwickshire) Navigation, Upper section
Authority: Upper Avon Navigation Trust.
Length: 20 miles 7 furlongs (34 km).
Locks: nine, 70 feet by 13 feet 6 inches (21 m by 4.1 m); draught 3 feet 6 inches (1.07 m).

In 1964 both the Southern Stratford-on-Avon Canal and the Lower Avon were reopened, but they were separated by an unnavigable section of the river Avon between Evesham and Stratford. Therefore it became important that the link should be restored to provide a through route. A fully navigable river Avon between Tewkesbury and Stratford would provide an important part of a circular cruising route, incorporating the river Avon, the river Severn and the Worcester and Birmingham and the Stratford Canals. Anonymous gifts of substantial amounts of money during the early stages of planning boosted morale and ensured an early start to the restoration project, so that work commenced in May 1969. Although it was to be a very difficult and expensive project, an expert in restoration work, Mr D. Hutchings, took over control. Previously he had managed the restoration of the Southern Stratford, where his hard work had made the project feasible. His

Old and new meet, as the motorway overshadows the Birmingham New Main Line.

expertise and efforts were fully stretched on the river Avon project, where the construction of nine new weirs was required, along with adjacent locks and canal sections. There was also the dredging of the whole route and the removal of many thousands of tons of rock and mud from the river bed to make a navigable channel. The efforts of the leader and the thousands of voluntary workers were rewarded when in June 1974 Queen Elizabeth the Queen Mother officially opened the river once more to navigation.

The waterway is a wonderful one, for the river runs through one of the most beautiful parts of England. Throughout the journey one can admire the beauty of nature and appreciate the physical efforts that were made by the dedicated enthusiasts who made the river navigable for all to explore.

Birmingham Canal Navigations
Length: 105 miles (169 km).
Locks: 139 (narrow-beamed).
Tunnels: Dudley, Netherton, Coseley, Curdworth, Ashted, Gosty Hill.

The network of canals in the Birmingham and Black Country district is basically situated at three different levels. They are the

Birmingham level at 453 feet above sea level (138 m), the Walsall level at 408 feet (124 m) and the Wolverhampton level at 473 feet (144 m). They are in predominantly urban surroundings, with numerous flights of locks, aqueducts and tunnels, based on a little over 100 miles (160 km) of canal. The system came about in 1768 with the first authorising Act for the Birmingham Canal Company to build a Canal from the Staffordshire and Worcester Canal at Aldersley to extend 22.5 miles (36 km) via twenty-nine locks into Birmingham. This waterway, engineered by Brindley, was opened throughout in 1772. Over the years the system was continually enlarged until by 1865 it extended well over 160 miles (257 km) and incorporated over two hundred locks.

Built to cater for the transport requirements of one of the world's largest industrial concentrations, by the end of the nineteenth century the network provided a means of moving nearly nine million tons of cargo each year. To move this vast tonnage, thousands of boats were in daily operation. The craft were loaded and discharged at numerous factories and business premises along the banks of the canal, as well as at over five hundred private basins.

The network that remains today passes through an interesting industrial area where there are waterway structures of architectural or engineering merit, though there are also many miles of pleasant scenery. As it provides a through route to connect a number of cruiseways, some sections are much busier than others. It takes a little over a week to explore the network by boat, although tow-paths exist alongside the canal and walking sections can be highly rewarding. Recent road development and land reclamation have cut back some branches, raising fears that little-used sections may still be threatened.

Brecon and Abergavenny Canal

Length: 35 miles (56 km).
Locks: six, 64 feet 9 inches by 9 feet 2 inches (19.7 m by 2.79 m). Acute bend near Llanfoist now limits passage of craft to 34 feet (10.4 m) in length.

This canal was designed by Thomas Dadford Junior and its history is closely linked with the tramroads that brought stone, iron and coal down to its banks. Its first Act of Parliament dates to 1793, when the canal was planned to run from Brecon to Caerleon, but because of the Monmouthshire Canal Company the plans were altered. So when it was opened in 1812 it ran from Brecon to Pontmoile, where it joined the Monmouthshire Canal. Finally in 1865 the two independent canal companies joined together and became the Monmouthshire and Brecon Canal Company. Because

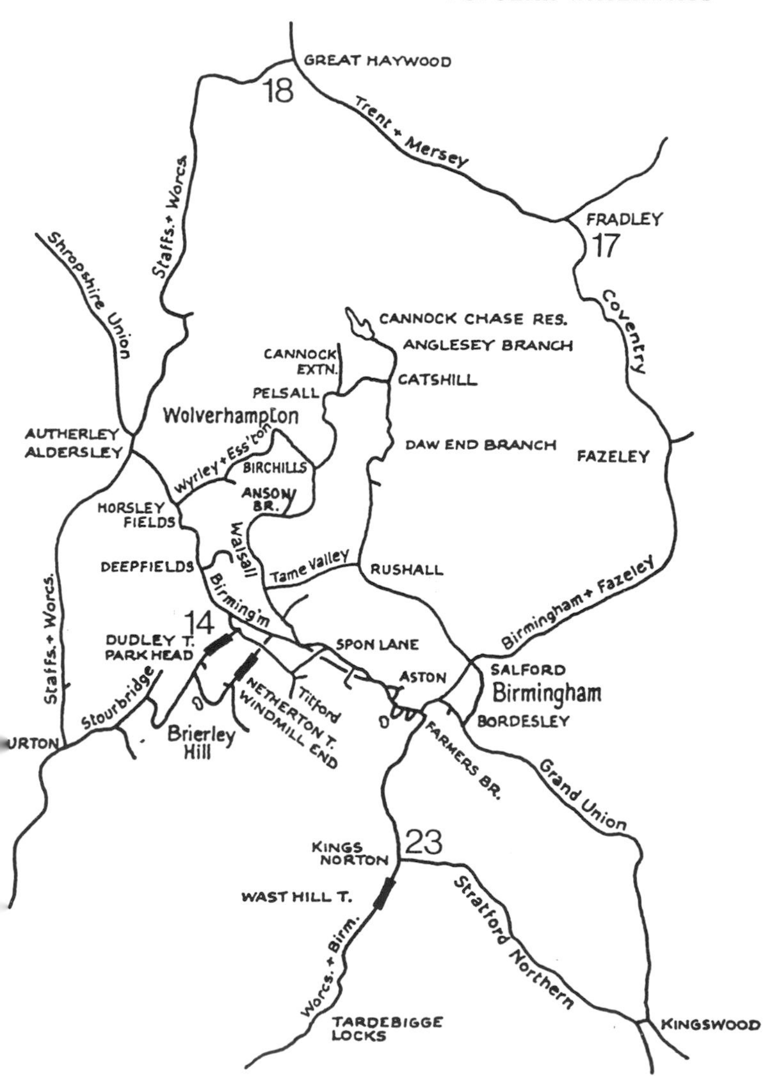

Map of the Birmingham Canal Navigations and connecting waterways. The numbers refer to places to visit described in Chapter 7: 14 Dudley Tunnel; 17 Fradley Junction; 18 Great Haywood; 23 Kings Norton Junction.

of railway competition, by the end of the nineteenth century there was very little traffic and the old Monmouthshire Canal was closed, the Brecon section being retained purely as a water channel.

In 1962 the network was officially abandoned, but the publicity drew attention to the fact that the canal ran through the Brecon Beacons National Park. Thus the potential of the canal was realised and work started on its restoration during 1964. A contour canal, the waterway clings to the sides of the mountains through the Usk valley. It is an isolated waterway, not connected to the rest of the canal system, but there is the most delightful scenery. It is navigable from Pontypool to Brecon, and, because of the existence of slipways, can be enjoyed by the owners of trailed boats as well as hirers of cruisers, which are available. Restoration to extend navigation below Pontypool is continuing.

Bridgewater Canal

Authority: Manchester Ship Canal (reciprocal arrangements exist on licences with the British Waterways Board).

Length: main line 23 miles 3 furlongs (37.6 km); branches 16 miles 3 furlongs (26.4 km).

Locks: one (Hulme Lock in Manchester, the connection with the Ship Canal; special arrangements are needed for its use).

The Rochdale Canal at Castlefield in Manchester.

On its eastern section the Bridgewater Canal runs for the most part through an industrial area, but on its western section it does provide some excellent cruising waters through peaceful countryside. The canal's main use now is for pleasure craft en route to other cruising waterways. Indeed it is an important link in the Cheshire ring of waterways, running from a junction with the Rochdale Canal in the city of Manchester to the Trent and Mersey Canal at Preston Brook. There is also a branch from there to Runcorn, where connection was once made with the Manchester Ship Canal. Within the built-up area of Salford is the Waters Meeting Junction, from where the original line of the canal runs through Worsley to Leigh. At Leigh it makes connection with the Leigh branch of the Leeds and Liverpool Canal, and this access to the canal provides further traffic, including many cruisers on their way to and from the Yorkshire waterways.

Calder and Hebble Navigation
Length: 21.5 miles (35 km).
Locks: twenty-seven, flood locks five, flood gates three; 57 feet 6 inches by 14 feet 3 inches (17.5 m by 4.3 m).

The navigation, originally engineered by John Smeaton, was opened from Wakefield to Salterhebble in 1770, and from there along the top pound to Sowerby Bridge in 1774. It is a typical navigation using the river for water supply and for part of its route. Commencing at Wakefield, it connects with the Aire and Calder Navigation, and halfway along, at Cooper Bridge, is a connection with the Huddersfield Broad Canal. At Sowerby Bridge, its terminal point, the connection with the Rochdale Canal, which was abandoned in 1952, has been reopened as the restoration of the Rochdale continues.

It was a busy waterway commercially until recently and it is classed as such for 9 miles (14 km) between Wakefield and Dewsbury. The rest, from Greenwood Lock upstream, is classed as a cruiseway. Barges carrying coal worked through the small locks to Dewsbury until the summer of 1981.

Environmental improvements, physical features with large locks, basins, a 200-year-old warehouse and historical connections make the waterway in Wakefield especially worthy of inspection.

A waterway that runs through the industrial areas of West Yorkshire is not expected to be beautiful, but there are some pleasant spots on the Navigation's commercial section and the upper region provides some surprises. The rolling hills of the Pennines blend with the solidly built structures of the waterway, and tree-lined sections that offer peace and tranquillity abound. West of Shepley Bridge, with its dry dock, are a number of single-storey stone-built lock keeper's cottages of a unique type and some company-built warehouses, some with wet docks.

Caldon Canal

Length: 20 miles 2 furlongs (32.6 km).

This canal joins the main line of the Trent and Mersey at Etruria and from there winds its way through some of the loveliest parts of Staffordshire. At Hazelhurst Junction, 8.5 miles (14 km) from the commencement of the canal, it divides, with the Froghall branch going by way of the beautiful Churnet valley, an area mostly inaccessible to road vehicles. At Froghall are the terminal basins where for many years stone from quarries at Cauldon Lowe was loaded into the boats. The Leek Branch from Hazelhurst Junction with a tunnel extends nearly three miles to the outskirts of Leek. The canal offers a fascinating experience for all who cruise along it.

Caledonian Canal

Length: 60 miles (97 km).

Locks: twenty-nine, 150 feet by 30 feet (45.7 m by 9.1 m), manned. Admiralty charts available —6 mph (9.6 km/h) speed restriction.

The Caledonian Canal, from Inverness to Corpach near Fort William, consists of canal sections linking the lochs Lochy, Oich, Ness and Douchfour. It was surveyed by John Rennie in 1793 and later, in 1801/2, by Thomas Telford, at the request of the government. They authorised the construction of the canal in 1803, with Telford as the chief engineer and with William Jessop as a consultant. It was opened to navigation in 1822. Many of the locks are grouped together in staircases and a total of twenty-nine locks had to be constructed, including three sea locks, together with eleven swing bridges over the canal. In recent years the locks have been modernised and the sea lock and basin at Corpach have been enlarged to take vessels up to 1,000 tons.

A coast-to-coast waterway, incorporating very large structures, it is basically a ship canal. Everything is of massive proportions, including the mountains alongside the lochs, the largest of which, Loch Ness, is 20.75 miles (33 km) in length. It traverses such beautiful scenery, with ruined castles and the mountains reaching upwards from the water's edge, that no other canal can compare with the Caledonian. This Scottish canal is popular with holidaymakers, and cruisers can be hired from several bases. There are also hotel boats and day trips.

Chesterfield Canal

Length: open section 26 miles (42 km).
Locks: sixteen (forty-nine on the derelict section); wide locks to Retford 72 feet by 13 feet 6 inches (21.9 m by 4.1 m) but bridge restrictions limit the width to less than this.

The canal connects at West Stockwith with the tidal river Trent, where the access lock is manned during normal working hours. The open section of the canal offers a peaceful and tranquil cruise through most pleasant countryside, particularly the section upstream from Drakeholes Tunnel (154 yards, 140 m) which passes through many wooded areas. Throughout the length of the canal, wildlife abounds: it seems that the pleasant winding waterway has more than its fair share of wild flowers, small animals and birds of many species.

The Act authorising construction of the canal was obtained in 1771 and it was opened in 1777. It is unusual in that, in a district of predominantly wide waterways, it has narrow locks on the top 20 miles (32 km), and this resulted in the operation of special narrow-beamed boats, which occasionally sailed on the tidal waters in the area. Traffic above Worksop ceased early in the twentieth century, and the canal was allowed to deteriorate when mining subsidence finally closed the Norwood Tunnel (3,102 yards, 2,836 m) between Worksop and Chesterfield. In 1968 the 26 miles (42 km) of canal

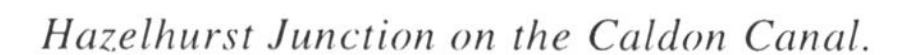

Hazelhurst Junction on the Caldon Canal.

Hartshill Maintenance Depot on the Coventry Canal.

from West Stockwith to Worksop was classed as a cruiseway, and the future for this open section is secure. Restoration of features and parts of the navigation is now proceeding in the section above Worksop.

Coventry Canal
Length: 38 miles (61 km).
Locks: thirteen (narrow-beamed).

The canal was authorised by Acts passed in 1768, 1786 and 1819 and was opened for traffic in 1790. It provided a through route from a junction with the Oxford Canal at Hawkesbury to the Trent and Mersey Canal at Fradley, with a 4.5 mile (7 km) section of canal from a terminal basin in Coventry to Hawkesbury Junction. These are still open, and along the route are connections with the Ashby Canal at Marston Junction, the Birmingham and Fazeley Canal at Fazeley Junction and the now closed extension of the Wyrley and Essington at Huddlesford. It once carried much coal traffic and continued with satisfactory trading figures and a regular dividend for the shareholders right up to nationalisation. It is not the pleasantest of canals in terms of scenery to cruise along, but it does have sections which pass through areas of natural beauty. It is a busy cruising waterway, for it connects with many other cruising waters both to the north and south and provides an alternative route to the Birmingham Canal Navigations.

Crinan Canal

Length: 9 miles (14.5 km).
Locks: fifteen, 88 feet by 20 feet (26.8 m by 6.1 m)

The canal was opened in 1809 after many engineering and financial problems. It is now used, as it always has been, as a short cut, to save a lengthy passage of 85 miles (137 km) around the Mull of Kintyre of vessels on passage to the Western Isles from Glasgow and the river Clyde. Used in the past by a great number of 'puffers', steam-powered cargo vessels, increasingly in recent years its role has been as a cruising waterway for sea-going yachts.

The canal runs from Ardrishaig to Crinan and at both locations there are large basins providing safe mooring away from tidal waters. On the 9 mile (14 km) route there are fifteen locks, including two sea locks, and, as expected in such an area, it has beautiful scenery all around.

River Derwent

Authority: National Rivers Authority.
Length: open section 22 miles (35 km).
Locks: two.

In 1701 the river was made navigable for 38 miles (61 km) inland from the river Ouse, and this was later extended by a further 11.5 miles (19 km) to Yedingham in 1805. The last section opened was the first part to be closed to navigation, while the river from Stamford Bridge to Malton was not used for navigation after the mid 1930s. Afterwards the river above Sutton Lock, which was not tidal, was gradually allowed to deteriorate and this too became unnavigable.

In 1972 Sutton Lock, the first lock on the river, 15.5 miles (25 km) inland from the Ouse, and closed in 1960, was reopened, when the Derwent Trust was established. This enabled craft to cruise along the 22 miles (35 km) of the river to Stamford Bridge. During the 1960s a water abstraction plant had been opened at Elvington and this was followed by the erection of a tidal barrage and lock at Barmby in 1975. The new structure at Barmby, built by the Yorkshire Water Authority, is a little downstream on the Ouse from the original confluence of waters, and a short section of channel had to be made to connect the lock with the original river, while the old connection was filled in. A licence is now required for boats wishing to cruise along the Derwent.

Running through an agricultural area, the river passes Wressle Castle alongside the A63 road, while several bridges of interest cross over the river. At Stamford Bridge the river is crossed by a mellow stone bridge dating from the 1720s, with the left-hand arch giving access to the now closed banana-shaped lock, where usually a boat will be moored. 11.25 miles (19 km) upstream from Barmby is the entrance to the Pocklington Canal.

Erewash Canal

Length: 11 miles 6 furlongs (18.9 km).

Locks: fourteen, 72 feet by 14 feet (21.9 m by 4.3 m).

The canal extends from Trent Lock on the river Trent near a junction with the Cranfleet Cut and runs in a northerly direction through Long Eaton to Langley Mill, where at the Great Northern terminal basin it connected with the now closed Cromford Canal. Two junctions en route connected with the now closed Derby Canal and the closed northern section of the Nottingham Canal. The waterway was originally opened in 1779 and was very busy for many years, but after the Second World War trade dwindled and for the most part the canal became derelict.

Because of its poor state, in 1968 it was classed as a 'remainder' waterway, but things were soon to change for restoration work was started by the Erewash Canal Preservation and Development Association, with the cooperation of the British Waterways Board and local authorities, with the result that it is now a fine open waterway. Though it runs mostly through an urban area, there is plenty of greenery, and it provides a pleasant journey with much to see. In

The disused Stamford Bridge Lock on the river Derwent.

A passenger boat about to pass through the Glory Hole in Lincoln.

the Long Eaton district a number of soundly constructed and imaginatively styled industrial premises can be seen and also a small number of well proportioned chimneys with elaborate finishing touches.

Fossdyke Canal
Length: 11.25 miles (18.1 km).
Locks: one, 74 feet 5 inches by 15 feet 2 inches (22.7 m by 4.6 m).

This is an ancient navigation, originally constructed by the Romans. It connects with the tidal river Trent by a lock at Torksey, passable by craft a little before and after high water. The lock is manned and usually operational at times that coincide with the tides during daylight hours, seven days a week. The canal provides a pleasant journey along a lock-free section to Lincoln, passing through the village of Saxilby about halfway. At Lincoln a swing bridge gives access to the large Brayford Pool, which, apart from the navigable channel, is now controlled by the local authority, and a payment is due for mooring, if only for a few minutes. At the north-east corner of the Brayford Pool is the narrow entrance to the Witham Navigation, a short distance along which is the famous Glory Hole. This is a bridge that carried both a road and buildings, the whole supported by brick-vaulted arches that can only be appreciated by those passing beneath them. From Lincoln the Witham Navigation continues for 31.5 miles (51 km) to the sea beyond the Grand Sluice Lock at Boston.

Grand Union Canal

Length: Brentford and junction with the river Thames to Birmingham 137.25 miles (221 km).

Lock limitations: London to Berkhamsted 77 feet by 14 feet 3 inches (23.5m by 4.3m), Berkhamsted to Braunston 72 feet by 12 feet 6 inches (21.9 m by 3.8 m).

In 1805 the Grand Junction Canal from Braunston on the Oxford Canal to Brentford in London was opened. Direct trade with London, using the Grand Junction Canal route, was only possible from Birmingham by using a number of other canals to Braunston. These included the Warwick and Napton Canal, which connected with the Oxford Canal near Braunston, and the Warwick and Birmingham Canal, which connected with the Digbeth Branch of the Birmingham and Fazeley Canal. Later, improved connection with the Birmingham network was possible when the Birmingham and Warwick Junction Canal was opened in 1842. This connected the Warwick and Birmingham Canal at Bordesley to the Birmingham and Fazeley and the Tame Valley Canal of the Birmingham Canal Navigations at Salford. These, briefly, were the waterway developments that went to make up the main line of the Grand Union Canal, which was formed in 1929.

The route between London and Birmingham had always been a busy one, and the new company was very enterprising and built up a large fleet of vessels while continually trying to improve its network. It obtained government support and made major improvements to the main line, including in the early 1930s the rebuilding and enlargements of locks. Despite all the efforts to make the Grand Union commercially viable, traffic gradually dwindled after the Second World War, and now the canal is used for the most part for pleasure cruising, but this activity keeps it very busy indeed.

The canal embraces everything that is best about inland waterways. During its construction the engineers and the navvies had enormous difficulties to overcome but that has resulted in one of the finest inland waterways in Britain, with numerous locks, tunnels, bridges and aqueducts. Many of these structures are marvels of engineering while the route, stretching over so great a distance, needs a book itself to describe the many changing facets of this cruiseway and its branches. In London access to the river Thames is possible and there are connections to other waterways in the area, including the River Lee Navigation via the Hertford Union

The ceremony of the blessing of the boats, held at Marsworth on the Grand Union Canal.

and Limehouse Cut from the Regent's Canal section of the Grand Union Canal. On the main line from Brentford to Braunston there are one hundred and two locks, and from Braunston to Birmingham there are sixty-four locks. The tunnels include Blisworth, Braunston and Shrewley.

Main line connections include:
The Regent's Canal, 8 miles 5 furlongs (13.9 km) with thirteen locks.
Hertford Union Canal, 1.25 miles (2 km) with three locks.
Paddington Arm, 13.5 miles (22 km).
Slough Arm, 4 miles 7.5 furlongs (8km).
Wendover Arm, 1.5 miles (2.4 km).
Aylesbury Arm, 6.25 miles (10 km) with sixteen locks.
Northampton Arm, 4 miles 6 furlongs (7.6 km) with seventeen locks.
Welford Arm, 1 mile 5 furlongs (2.6 km) with one lock.
Leicester Line, from Norton Junction to the river Trent (see below).

Grand Union Canal (Leicester Line)
Length: 66 miles 1 furlong (106 km).
Locks: fifty-nine, seventeen of them between Watford and Foxton inclusive (narrow-beamed), the rest to the river Trent 72 feet by 14

feet (21.9 m by 4.3 m).

The waterway from the Grand Union main line at Norton Junction extends via Crick tunnel (1,527.5 yards or 1,397 m), Husbands Bosworth tunnel (1,170.75 yards or 1,071 m), Foxton, Saddington tunnel (881.75 yards or 809 m), Leicester and Loughborough to a junction with the river Trent at Redhill Lock. It consisted originally of a number of independent waterways that were absorbed in 1932 by the Grand Union Canal Company. The oldest part of the route is the former Loughborough Navigation from the river Trent to Loughborough, based on the river Soar. The section from Loughborough to Leicester followed in 1794. Both these sections were built to accommodate wide-beamed craft. A further extension of the waterway inland was by the Leicester and Northampton Union Canal, authorised in 1793. By 1797 this had only reached as far as Debdale, north of Foxton, because of financial problems, but fortunately a group of promoters was able to complete the remaining section of canal through to Norton Junction by 1814, but only with narrow-beamed locks.

The first part of the route south from the river Trent makes use of the river Soar as a navigation to Leicester. It is a very attractive waterway, with beautiful scenery and an abundance of wildlife, and it passes through or near to many interesting villages. The

The 'Tunbridge' lifting bridge on the Huddersfield Broad Canal.

route through Leicester is interesting. In periods of heavy rain, navigation of the river can be dangerous and is not recommended for the inexperienced. In Leicester, in particular, flooded towpaths can hide the exact channel, while high water restricts passage through certain bridge holes. There is the further danger of swift moving water flowing over weirs, as on all navigations. From Leicester it is purely canal, with renowned waterway features such as Foxton and the several tunnels.

Grand Western Canal

Authority: Devon County Council, Exeter.
Length: 11.25 miles (18.1 km).
Locks: none.

In 1796 an Act was obtained for the construction of this canal, to run from Topsham to Taunton, with three branches. At Taunton connection would be made with the rivers Tone and Parrot. Owing to inflation due to the Napoleonic war, work stopped until 1810, when John Rennie became involved. Then work recommenced on the 11.25 mile (18 km) section from Lowdwell to Tiverton and this was completed in 1814 at a cost of £220,000. During 1827 work started on the Taunton to Lowdwell section and this was completed in 1838. It had several vertical lifts and an inclined plane but all other sections of the original scheme were abandoned. In 1854 the railway took control of the canal, the minimum of maintenance was carried out and traffic was gradually reduced until 1924, when a serious breach cut the canal into two separate sections.

In 1971 the original 11.25 mile (18 km) section of canal was taken over by the Devon County Council for recreational use. They established a linear canal walk, which is now well used, and this provides an opportunity to see the waterway. The exploration of the canal by water is not as easy, for only non-motorised craft are allowed and overnight mooring is not permitted. A horse-drawn passenger boat operates from Tiverton, and from there rowing boats can be hired. A number of canal artifacts, along with bridges and industrial remains, can be seen on this interesting isolated waterway.

Huddersfield Broad Canal

Length: 3.7 miles (6 km).
Locks: nine, 57 feet 6 inches by 14 feet 2 inches (17.5 m by 4.3 m).

The canal is nearly 3.75 miles (6 km) long with the Aspley Basin providing many facilities near its town centre end, where the Broad Canal connects with the Huddersfield Narrow Canal which is now being restored. The Huddersfield Broad Canal was opened in 1776 and was built to connect with the Calder and Hebble Navigation at Cooper Bridge. The canal was dominated by railway ownership until 1944 when it was sold to the Calder and Hebble Navigation for £4,000. It was never a very successful waterway, but it did manage to keep some regular traffic until the mid 1950s, after which it deteriorated quickly until 1968, when it was given 'cruiseway' status. This provided the stimulus to carry out maintenance work and a practically unnavigable waterway was transformed, for since then it has seen increasing numbers of pleasure craft.

The scenery along this short wide canal is mixed, ranging from open spaces to busy industrial areas. Alongside most of the locks is a stone-built bridge of a distinguished style. Towards the end of the canal, well within the town, is a unique example of a lifting bridge. This was originally erected in 1865 to replace a former swing bridge. It was rebuilt a few years ago, but the Turnbridge lost nothing of its character, with an overhead structure of girders holding a number of large wheels, and the deck, which is suspended below, moving up and down parallel to the water.

Kennet and Avon Canal

Length: 86.5 miles (139 km).
Locks: 106, 67 feet by 14 feet (20.4 m by 4.3 m).

The canal runs from a junction with the river Thames at Reading to the river Avon (Bristol) at Hanham. Altogether it took seven different Acts of Parliament to bring it into being, the first of them dating from 1794. The waterway consists of the former River Avon Navigation, the River Kennet Navigation and the Kennet and Avon Canal. Shortly before nationalisation the waterway became rather neglected and as a result it was no longer navigable throughout by the early 1950s. But one of Britain's most ambitious waterway restoration schemes was started and, after the culmination of many years hard work by the Kennet and Avon Trust, and the many volunteers as well as the British Waterways, the canal was officially re-opened early in August 1990 by Her Majesty the Queen. Back pumping schemes are increasingly overcoming the previous problem of water shortages which has restricted navigation.

It is a wonderful waterway, with some very fine scenery on its route through the delightful countryside. The canal is renowned for its great engineering works, ranging from the locks at Devizes to the Dundas Aqueduct, the Avoncliffe Aqueduct, the Claverton and Crofton Pumping Stations and the 502 yard long (459m) Bruce Tunnel.

Lancaster Canal

Length: 42.5 miles (68 km).
Locks: seven on the Glasson Branch, 72 feet by 14 feet 6 inches (21.9m by 4.4 m).

John Rennie surveyed the proposed 75 mile long (121 km) route from Easthoughton, near Wigan, through Preston and Lancaster to Kendal. It was intended to include locks to take the canal the 272 feet (73 m) from its levels in Lancashire down into the Ribble valley, and up the 72 feet (22 m) from Tewitfield to Kendal. There were also to be two large aqueducts to cross the rivers Ribble and Lune. Parliamentary authorisation was given in 1792 and work began on making the new waterway. First of all, they made the canal at two separate sites, one on each side of the river Ribble. There, instead of building an aqueduct, they built a wooden bridge to carry a tramroad 5 miles (8 km) in length, which eventually joined the first two sections of canal together. From such unusual beginnings they eventually constructed a canal from Kendal, via the Hincaster Tunnel and eight locks at Tewitfield, and carried over the river Lune by a very fine aqueduct, through Lancaster to Preston, with a branch to Glasson Dock. The original section of canal south of the Ribble and isolated from the rest they sold to the Leeds and Liverpool Canal.

Beautiful stonework on the Kennet and Avon Canal at Bath.

The section of canal from Tewitfield to Kendal was unfortunately cut off from the rest of the canal when the M6 motorway was built. 2 miles (3.2 km) of the canal within Preston have also been closed, and this leaves an isolated waterway with a connection to the sea via Glasson Dock. Nevertheless the canal has its admirers, for it runs through beautiful scenery with rocky cuttings, splendid views of Morecambe Bay and the surrounding areas.

Lee and Stort Navigation
Length: Lee, 27.75 miles (45 km); Stort, 13.25 (21.3 km).
Locks: twenty-one Limehouse to Harlow, 88 feet by 15 feet (26.8 m by 4.6 m); nineteen Harlow to Old Ford Lock, 88 feet by 13 feet 4 inches (26.8 m by 4.1 m).

The river Lee was one of the earliest river navigations, dating from Roman times, but it is not used commercially today. The first Act of Parliament to improve the navigable channel was in 1425, but it was subject to other acts from 1739. In 1868 the Lee Conservancy Board took over control of the navigation, and during the First World War the navigation was improved to take 130 ton capacity vessels to Enfield, which later necessitated the duplication and mechanisation of some of the locks. At the time of nationalisation, the navigation passed to the state-controlled British Waterways.

Attempts were made to make the river Stort navigable after an Act in 1759, but unfortunately lack of financial support then and afterwards continued to delay the completion of a satisfactory waterway. The waterway seems to have been a continual problem for all concerned with it throughout the years, and in 1911, simply to get rid of the problems associated with it, one company sold it for 5 shillings (25p). Finally, in the early 1920s it was completed.

The Lee section has connections with the river Thames at Limehouse and via Bow Creek at Blackwall. It extends north to Harlow Mill and beyond to Hertford, with the Stort section making connection at a junction at Harlow Mill and extending from there to Bishop's Stortford. The network includes the 1.75 mile (2.8 km) branch to the Regent's Canal via the Hertford Union Canal. It is a typical navigation, with both canal and river sections. The lower part of the river Lee passes through a busy industrial area, while the upper sections, especially of the Stort, are not affected by industry and have some very pleasant stretches for cruising.

Leeds and Liverpool Canal
Length: 126.5 miles (204 km) (main line).
Locks: ninety-two, east of Wigan 62 feet by 14 feet 4 inches (18.9 m by 4.4 m); west of Wigan 72 feet by 14 feet 4 inches (21.9 m by 4.4 m).

The double bridge at East Marton on the Leeds and Liverpool Canal.

In 1720 an Act was passed to make the river Douglas navigable inland from the river Ribble to Wigan. Then in 1770 an Act was passed to make a canal all the way from Leeds to Liverpool. When it was originally surveyed by John Longbottom it was planned to be 108.75 miles long (175 km), but when it was finally completed in 1816 the main line of the canal totalled 126.5 miles (204 km). This included 7 miles (11 km) of the old Douglas Navigation and an isolated part of the Lancaster Canal. A connection from Wigan to Manchester was opened in 1821, this branch joining the Bridgewater Canal at Leigh.

The canal, with ninety-one locks between Leeds and Liverpool, has a summit pound 487 feet (148 m) above sea level. To achieve this height over the Pennines required some exceptionally heavy engineering work and needed numerous aqueducts and two tunnels one at Foulridge and a smaller one at Burnley. Also at Burnley there is a 1,256 yards (1,148m) embankment 60 feet (18 m) high. The locks in particular are massive structures, especially the staircase locks, with the Bingley Five Rise the most impressive of them, and the Wigan Flight of twenty-one locks also very fine.

Pleasure craft use the canal more and more, so that certain sections are now busy, but generally the canal offers peace and tranquillity. The canal passes through some magnificent moorland scenery and it is a wonderful experience to cruise in the Pennine areas. A paradise for walkers, the canal offers special appeal throughout the four seasons. Even the lengths of canal within the industrial areas are not without interest, for there is much to see, including canal structures and fine buildings of days gone by.

Llangollen Canal

Length: 46 miles (74 km).
Locks: twenty-one (narrow-beamed).

This canal, sometimes referred to as the Welsh Canal, was for many years an important part of the Shropshire Union Canal system, which obtained its first Act in 1793. The major part of the present canal was authorised by an Act in 1801, and this was followed by a further one that authorised a navigable feeder from the river Dee at Llantysilio, near Llangollen, to the end of the Pontcysyllte Aqueduct.

The Llangollen Canal runs from Llantysilio, 1.5 miles (2.4 km) west of Llangollen, to the Shropshire Union Canal at Hurleston Junction, near Nantwich. The canal is exceedingly picturesque, with many great feats of engineering. It is a very attractive waterway throughout, from the Horseshoe Falls at the head of the canal, to its bottom lock at Hurleston. There are three tunnels, Ellesmere (87 yards or 80 m), Whitehouse (191 yards or 175 m),

A lock keeper's cottage at Grindley Brook on the Llangollen Canal.

A stone bridge below a lock on the Macclesfield Canal.

and Chirk (459 yards or 420 m), and two very famous aqueducts, Chirk and Pontycysyllte, along with several junctions.

Macclesfield Canal

Length: 26.5 miles (43 km).

Locks: thirteen (narrow-beamed).

The Macclesfield Canal connects at Harding's Wood Junction, Kidsgrove, with the summit level of the Trent and Mersey Canal and from there passes through Congleton, Macclesfield and Bollington on its way to Marple, where it connects with the Peak Forest Canal. Originally constructed under an Act of 1826, it was opened in 1831.

In the 1960s, because it no longer offered a through route, it was a very quiet waterway, where it was possible to cruise for several days without seeing another boat on the move. Now it is very different, for since the restoration of the Peak Forest and Ashton Canals it has become part of the Cheshire ring of cruising waterways and it is very busy. It passes through beautiful rolling countryside that delights most people. There are numerous well proportioned mellow stone accommodation bridges, and the Bosley flight of twelve locks, fitted with double mitre gates at each end. It is one of the best canals to cruise along. To moor and explore is no waste of time, for along the whole length of the canal there is much to be seen, from Tudor buildings to leafy country lanes.

Milton Ferry bridge on the river Nene.

River Nene
Authority: National Rivers Authority.
Length: 65 miles 5 furlongs (106 km).
Locks: thirty-seven, 78 feet by 13 feet (23.8 m by 4.0 m).

Commencing at Northampton, with a connecting branch from the Grand Union Canal, the river route extends to the Dog in a Doublet Lock, 5 miles (8.0 km) below Peterborough. Beyond this the river is tidal to the port of Wisbech and the sea. At Peterborough there is a junction with a branch waterway that leads to the Stanground and Middle Level Navigations. It is an old river navigation, whose first Act of Parliament dates from 1714. A most attractive waterway to explore by boat, it runs through meadows and skirts many pleasant little villages worthy of further exploration, and at many places by the side of the river are old mills. The locks were improved in the 1930s and most are fitted with a guillotine gate at the bottom. When fresh water is flowing it is normal to find excess water flowing over the top wooden gates, and on such occasions extra care must be exercised.

Oxford Canal
Length: 77 miles (124 km).
Locks: forty-six (narrow-beamed).

The first Act for a canal from Longford near Coventry to Oxford received the royal assent in 1760. Construction began at the

Longford end, but when only 16 miles (26 km) had been completed, the engineer, James Brindley, died and his assistant, Samuel Simcock, took over responsibility for the work on the canal. Steady progress was maintained and it was completed as far as Napton in 1775 and Banbury in 1778. Then work stopped for a while, but finally the canal was completed to Oxford, where connection was made with the river Thames in 1790. During 1805 the Grand Junction Canal to London was opened and the section of the Oxford south of Napton lost a great deal of traffic to the new, more direct route to London. However, the northern section remained a busy waterway.

A twisting and turning canal because of the contours, its overall length was reduced by 13.5 miles (22 km) by new cuts on the northern section, constructed by Telford between 1829 and 1834. Commencing at the Sutton Stop Lock at Hawkesbury Junction, where it connects with the Coventry Canal, the Oxford Canal runs through some wonderful countryside on the way to Braunston. There the Grand Union Canal and the Oxford Canal share the same route for a short distance to Napton Junction, from where the Oxford Canal continues through charming countryside, with many interesting canalside buildings and delightful lift bridges. It is now a very popular cruising waterway, and this results in water short-

A narrow lock in beautiful surroundings on the charming Oxford Canal.

ages during most summers, on the southern section in particular.

Peak Forest Canal

Length: 14.75 miles (23.7 km).
Locks: sixteen (narrow-beamed).

This canal was authorised by an Act of Parliament in 1826. The canal's terminal basin is at Whaley Bridge, where there is an excellent stone-built warehouse. From there the canal follows the Goyt valley through to Marple, where, by a flight of sixteen locks, it descends to cross the river Etherow by the Marple Aqueduct. This is followed by Hyde Bank and Woodley tunnels, to connect with the Ashton Canal at Dukinfield.

A beautiful canal, it follows a tree-lined winding course for much of its way. Half a mile (800 m) from Whaley Bridge Basin there is a junction, and from there a short arm leads to Buxworth Basin which was once the head of navigation. From the terminal point to the junction with the Macclesfield Canal at Marple the canal remained open when the rest became derelict in the early 1960s. The part of the canal that was out of use for many years was restored and reopened in the mid 1970s.

Pocklington Canal

Length: 9.5 miles (15.3 km).
Locks: nine, 57 feet by 14 feet 3 inches (17.4 m by 4.3 m), not all open.

The Pocklington Canal was opened in 1818 and extended 9.5 miles (15.3 km) from the river Derwent at East Cottingworth, with nine locks, to a terminal basin situated alongside what is now the A1079 road. Built to serve an agricultural area, it was never very profitable and was taken over by the York and North Midland Railway in 1847. The last commercial use of the waterway was in 1932 and after that it became derelict. During 1962 the swing bridges were replaced by fixed low level units which stopped any attempt to cruise along the canal. Concern for the future of the waterway resulted in the formation of the Pocklington Canal Amenity Association. Helped by the British Waterways Board and with financial support from local councils, the members have now reopened half of the waterway. The open section now extends from the river Derwent to Melbourne.

Ripon Canal and connections

Locks: 58 feet by 14 feet 6 inches (17.7 m by 4.4 m).

This is the most northern canal in England that is connected to the major system of inland waterways. The canal received the royal assent authorising its construction in 1767. The Act ap-

Tunnel under the canal especially for horses at Marple on the Peak Forest Canal.

proved the making navigable of part of the river Ure from a junction with the river Swale and through Boroughbridge to Ox Close, from where a canal section continued the waterway to Ripon. Completed in 1773, the canal section, with an entrance lock at Ox Close and two more along its route, extended 2.25 miles (3.6 km) into Ripon. In 1955 the two top locks were cascaded, but the entrance lock and the first 1.25 mile (2 km) pound remained to serve the members of the Ripon Boat Club. During 1987 both the closed locks were reopened.

The canal is accessible from York, by a journey along the Ouse, through Linton Lock and to Ouse Gill Beck, the confluence of the rivers Ure and Swale, then along 8 miles (12.9 km) of the navigable river Ure, which includes two short canal cuts, with locks at Milby and Westwick. The Milby lock is on the outskirts of Boroughbridge, at the end of a 0.5 mile (0.8 km) cut, while Westwick lock is located 5 miles (8.0 km) along the river. The whole waterway is beautiful and, even in these days of appreciation of inland waterways, is very much underused.

Selby Canal

Length: 11.75 miles (18.9 km).
Locks: four, 78 feet 6 inches by 16 feet 6 inches (23.9 m by 5.0m).

The Selby Canal, approved in 1774, was built by William Jessop, the engineer for the Aire and Calder Navigation. For many years it was part of that waterway's main line connection with the tideway. Opened in April 1778, it extends inland for 5.25 miles (8.4 km) from the Selby Lock junction with the river Ouse to Haddesley Lock. From there the river Aire is used for the 6.5 mile journey (10.5 km) to Bank Dole Lock, Knottingley, where it connects with the main line of the Aire and Calder Navigation.

The river section, with high flood banks, follows a very twisting course with one lock, at Beal, 2 miles (3.2 km) from Knottingley. The canal section, with a peaceful charm of its own, cannot offer any remarkable features, natural or otherwise, but it has an abundance of wildlife, including many species of birds.

River Severn

Length: 42 miles (68 km) Gloucester to Stourport.
Locks: five, Gloucester to Worcester 135 feet by 22 feet (41.1 m by 6.7 m), Worcester to Stourport 89 feet by 18 feet 11 inches (27.1 m by 5.8 m).

The River Severn was one of Britain's earliest and most successful river navigations. The Acts of Parliament for its improvement date from 1503. Originally it was navigable from the sea inland to Shrewsbury and beyond, on high water, to Welshpool but now it is

Gloucester Docks.

navigable regularly only as far as Stourport. A scheme exists to extend locks upwards to Shrewsbury and link the river with the Shropshire Union canals.

From the Bristol Channel traffic gains access to Gloucester via the Gloucester and Sharpness Canal, and from there uses the river Severn, bypassing weirs by modernised electric locks. The towns and cities along the course of the river add interest for the visitor. They include Gloucester, Tewkesbury, Worcester and Stourport. In addition to the Staffordshire and Worcester Canal connecting with the river at Stourport, access can also be gained to the Worcester and Birmingham Canal at Worcester and the river Avon at Tewkesbury and, it is hoped, in the near future with the restored Droitwich Canal a little below Stourport. In recent years there has been a big increase in the number of pleasure craft using the river and they are catered for by a number of new marinas. The National Waterways Museum in Gloucester Docks is a leading example of the reclamation of old buildings and a must for anyone with an interest in inland navigation.

Sheffield and South Yorkshire Navigation

Length: 42.75 miles (69 km).
Locks: thirty-one.

The Sheffield and South Yorkshire Navigation, which incorporates the Sheffield Canal and the Stainforth and Keadby Canal, extends over 42.75 miles (69 km) from Keadby and the river Trent inland to the centre of Sheffield. In 1726 an Act was passed

A pair of hotel boats on the Shropshire Union Canal.

allowing for improvements to the river Don so that vessels could reach the outskirts of Sheffield, while in 1819 the Sheffield Canal, 4 miles (6.4 km) in length, was opened, so connecting the old navigation with the centre of Sheffield by water. The Sheffield and South Yorkshire Navigation was then formed in 1888. The new company, with a capital of £1,500,000, acquired the three waterways between the Trent and Sheffield, as well as the Dearne and Dove Canal, which is now closed.

The lower section of the waterway below Rotherham was classified as a commercial waterway in 1968. After many years of hesitation by successive governments, during 1978 money was allocated for the improvement of the commercial section, from the New Junction Canal (which also had improvement work undertaken) upstream to Rotherham. The improvement of work completed in 1983 enables craft with a cargo capacity of up to 700 tons to operate to and from the Humber ports by use of the adjoining waterways and Goole. Commercial traffic operations are spasmodic.

The scenery along the waterway ranges from level agricultural areas between Keadby and the town of Thorne, a popular pleasure boat centre, to straggling urban areas upon hillsides. The stretch upstream from Doncaster has a special appeal with many pleasant landscaped views and beautiful wooded river sections. Some have been transformed as a result of environmental improvements. Large electrically operated locks give way above Rotherham to the original small 61 feet 6 inch long (18.7 m) hand-operated locks. These manual locks include the eleven of the Tinsley Flight on the old Sheffield Canal section that leads to the city basin, currently being developed.

Shropshire Union Canal

Length: 66.5 miles (107 km) plus 10 mile (16.1 km) Middlewich Branch.
Locks: forty-six, plus four on Middlewich Branch, 70 feet by 13 feet (21.3 m by 4.0 m) from Nantwich to Ellesmere Port, elsewhere narrow-beamed.

In 1772 an Act was passed for a wide-beamed canal from Chester to Nantwich, followed twenty-one years later by another Act for a canal from Ellesmere Port to Chester. These two now make up the northern part of the Shropshire Union Canal. In 1826 the Birmingham and Liverpool Junction Canal Act allowed for the building of the section of canal between Nantwich and its southern end at Authersley Junction. There connection is made with the Staffordshire and Worcester Canal. Later a branch canal was authorised from Barbridge, north of Nantwich, to Middlewich to connect with the Trent and Mersey Canal. They were all grouped together under the Shropshire Union Canal Company Act of 1845, and that also included the Llangollen Canal from Hurleston Junction.

The canal runs in a northerly direction from Authersley Junction, near Wolverhampton, to Ellesmere Port via Chester. At Ellesmere Port, where it connects with the Manchester Ship Canal, the old docks are enjoying a revival based around the Boat Museum with its magnificent collection of inland waterway craft.

Staffordshire and Worcester Canal

Length: 46.5 miles (75 km).
Locks: forty-three (narrow-beamed).

Engineered by James Brindley, the canal is a contour canal of the old pattern. It runs from a junction with the Trent and Mersey Canal at Great Haywood to the river Severn at Stourport. Opened in 1772, it became a tremendous success, paying high dividends to the shareholders for many years.

It is a pleasant cruising waterway, the southern section being particularly attractive, with a unique type of circular overflow weir alongside many of the locks. In many places there are deep narrow

The canal wharf at Tyrley on the Shropshire Union Canal.

cuttings through red sandstone and also a number of short tunnels. Despite being so near the industrial Midlands, the canal winds its way through beautiful scenery and through many wooded sections, retaining an old world charm with its many little brick-built bridges and pretty lockside cottages. The canal connects with the following cruising waterways: the Severn at Stourport, the Stourbridge Canal, which leads to the Dudley and Netherton Tunnels, the Birmingham Canal Navigations at Aldersley Junction and the Trent and Mersey Canal. With so many connections it is inevitably a busy waterway, but its numerous users include many who admire it for its own sake.

Stratford-on-Avon Canal (Northern section)
Length: 12 miles 4 furlongs (20.1 km).
Locks: twenty (narrow-beamed).

The canal extends from the Worcester and Birmingham Canal at Kings Norton Junction to Kingswood Junction on the Grand Union Canal, only a few hundred yards beyond a junction with the Southern section of the Stratford Canal. At Kings Norton Junction there is a guillotine stop lock, and from there a 10.5 mile pound (16.9 km) leads to the Lapworth Flight, which leads down to the junctions. 1 mile (1.6 km) from Kings Norton is a tunnel, 352 yards long (322 km).

Although so near to Birmingham, the canal seems to be far from urban influence as it runs through very pleasant country area. The surroundings are rich in historical associations, ranging from Roman remains to moated mansions, some of which are the property of the National Trust.

Stratford-on-Avon Canal (Southern section)
Length: 13 miles 4 furlongs (21.7 km).
Locks: thirty-six (narrow-beamed).

The Southern Stratford extends from the junction with the Northern Stratford Canal to Stratford-upon-Avon. There, at the basin end, is a wide lock and access to the river Avon. The authorising Act was passed in 1793, and the canal as a whole, including the Northern and Southern sections, was opened in 1816.

The canal was closed to navigation for many years, and in 1958 attempts were made to abandon it officially. This retrograde step was only stopped as a result of a campaign by the Inland Waterways Association and other groups. In 1960 the National Trust took over the responsibility for the canal, and in April of the following year restoration work began at Kingswood. The work, undertaken by volunteers, servicemen and prisoners, included repairs and rebuilding work on locks and bridges, walls and aque-

A motorboat enters the top lock of the Bratch Locks on the Staffordshire and Worcester Canal.

A lock keeper's cottage on the Stratford-on-Avon Canal.

ducts, as well as extensive dredging and towpath restoration. The success of their efforts was acclaimed when the canal was ceremonially reopened in July 1964. On 1st April 1988 formal control of the canal passed from the Trust to British Waterways.

On the canal there are many split bridges, which saved unnecessary unhitching of the towing horses. Beside a few of the locks are brick-built cottages with barrel-shaped roofs, which look similar in shape to a short section of canal tunnel with end walls. The canal, running through the very heart of England, is charming and the many villages on the route add further interest.

Trent and Mersey Canal
Length: 93.5 miles (150 km).
Locks: seventy-five, 70 feet by 13 feet 6 inches (21.3 m by 4.1 m) from the river Trent at Derwent Mouth to Horninglow Wharf, Burton-on-Trent; elsewhere, except for two at the western end of the canal, they are narrow-beamed.

A cross-country canal originally known as the Grand Trunk Canal, this is one of Brindley's masterpieces, although he never lived to see it completed, as he died in 1772. Eleven different Acts of Parliament were needed for the construction of this waterway. It has features of every kind, from long tunnels to cuttings and a boat lift. The tunnels are at Preston Brook (1,239 yards or 1,133 m), Saltersford (424 yards or 338 m), Barnton (572 yards or 523 m), and Harecastle (2,919 yards or 2,669 m). There are seventy-six locks, many of which had to be duplicated to accommodate the large number of trading vessels it once had. They number 1 to 40 from Derwent Mouth, rising to the summit level, and 41 to 75, falling to Middlewich.

The canal connects with the river Trent at Derwent Mouth, the Coventry Canal at Fradley Junction, the Caldon Canal at Etruria, the Staffordshire and Worcester Canal at Great Haywood, the Macclesfield Canal at Hardings Wood Junction, the Middlewich Branch of the Shropshire Union Canal at Middlewich, the river Weaver at Anderton via the boat lift, and the Bridgewater Canal at Preston Brook. With a main line of over 90 miles (145 km), there is a wide range of canalside scenery, from a collection of power station chimneys to almost an aerial view of a section of the river Weaver. It passes through charming and picturesque villages but also some grimy-looking towns. Impressive warehouses, pottery bottle kilns and tree-lined avenues can be admired from this popular cruising waterway, which provides peace and contentment in one section and the noise of busy railway lines in another.

River Weaver

Length: 20 miles (32 km).

Locks: five, 130 feet by 35 feet (39.6 m by 10.7 m), Northwich to Winsford; 150 feet by 35 feet (45.7 m by 10.7 m), Northwich to

The Trent and Mersey Canal from Etruria.

The Worcester Commandery and the Worcester and Birmingham Canal.

Weston Point.

This is an old river navigation which extends inland from Weston Point Docks and the access lock from the Manchester Ship Canal to Winsford. On the waterway there are five manned large capacity locks and a number of large swing bridges that allow small coasters to operate upstream 13 miles (21 km) to Northwich. The usual access point to the waterway for pleasure craft is at Anderton, where they are lowered by the boat lift from the Trent and Mersey Canal.

On the first part of the journey from Weston Point the banks are covered by great expanses of chemical works, but thereafter the route is through pleasant rural areas until Northwich. There, opposite the boat lift, is another large concentration of chemical works. Upstream to Winsford it is very pleasant for the river is wide and deep and mostly the banks are lined with trees.

Worcester and Birmingham Canal

Length: 30 miles (48 km).

Locks: fifty-eight; first two at Worcester 76 feet by 18 feet 6 inches (23.1 m by 5.6 m); others narrow-beamed.

In the year 1791 the first Act was passed for the construction of the Worcester and Birmingham Canal, to run, as the name implies, from Worcester to Birmingham. At Diglis, Worcester, a large canal basin was made, two locks up from the river Severn, to allow the sailing vessels of the river, the trows, to gain access to tranship

The Worcester and Birmingham Canal at Shernal Green.

cargoes. At the other end of the canal, in Birmingham, physical connection at Worcester Bar with the Birmingham Canal Navigations was banned. As a result, for a while all goods had to be carried over an intervening strip of land to be reloaded into different boats. This unusual practice continued until 1815, when a stop lock connection was agreed upon.

It was a very difficult and expensive canal to build, testing the ingenuity and skill of all concerned, for they had to lift the canal over 400 feet (112 m) up to the Birmingham level through rolling hills. So the canal, for its length, required a comparatively large number of cuttings and tunnels to keep the number of locks required to a minimum. Even then they had to spend a great deal on lock construction, for fifty-eight locks were required in 15 miles (24 km) of canal. These included the greatest of all flights, the Tardebigge Flight of thirty locks within a 2.5 mile (4.0 km) section of canal.

The tunnels on the canal are Dunhampstead (230 yards or 210 m), Tardebigge (580 yards or 530 m), Shortwood (613 yards or 561 m), West Hill (2,726 yards or 2,493 m), and Edgbaston (105 yards or 96 m). At Kings Norton, 5.5 miles (8.9 km) from the Worcester Bar terminal, is a junction with the Stratford on Avon Canal. Former connections with other waterways existed at Hanbury with the Droitwich Junction Canal and at Selly Oak with the Dudley Number 2 Canal. Apart from the last 5 miles (8.0 km) into Birmingham, the canal is through very pleasant country, which includes a fruit growing area. The canal provides a good deal of physical exercise but healthy enthusiastic crews who are young in spirit will consider it a must.

Swing bridge over the Droitwich Canal, Worcestershire.

9. Museums

Museums devoted entirely to inland waterways are denoted by an asterisk*.

The Black Country Museum, Tipton Road, Dudley, West Midlands DY1 4SQ. Telephone: 0121-557 9643.

*The Boat Museum, Ellesmere Port, Cheshire L65 4PW. Telephone: 0151-355 5017.

*The Canal Museum, Canal Street, Nottingham. Telephone: 0115-959 8835.

*The Canal Museum, LUCS, Manse Road Basin, Linlithgow, West Lothian. Telephone: 01506 671215.

*The Canal Museum, The Sobriety Centre, Dutch Riverside, Goole DN14 5TB. Telephone: 01405 768730.

*Canal Museum, Stoke Bruerne, Towcester, Northamptonshire NN12 7SE. Telephone: 01604 862229.

*Dolphin Sailing Barge Museum, Crown Quay Lane, Sittingbourne, Kent ME10 3SN. Telephone: 01795 423215.

Exeter Maritime Museum, The Haven, Exeter, Devon EX2 8DT. Telephone: 01392 58075.

Ironbridge Gorge Museum, Ironbridge, Telford, Shropshire TF8 7AW. Telephone: 01952 433522.

Lancaster Maritime Museum, St George's Quay, Lancaster LA1 1RB. Telephone: 01524 64637.

*London Canal Museum, 12-13 New Wharf Road, London N1 9RT. Telephone: 0171-713 0836.

Merseyside Maritime Museum, Albert Dock, Liverpool L3 4AQ. Telephone: 0151-207 0001.

Morwellham Quay Museum, Morwellham, Tavistock, Devon PL19 8JL. Telephone: 01822 832766.

Museum of Technology, Cheddars Lane, Cambridge. Telephone: 01223 368650.

*National Waterways Museum, Llanthony Warehouse, The Docks, Gloucester GL1 2EH. Telephone: 01452 318054.

North Devon Maritime Museum, Odun Road, Appledore, Devon EX39 1PT. Telephone: 01237 422064.

Town Docks Museum, Queen Victoria Square, Hull HU1 3DX. Telephone: 01482 593902.

Welsh Industrial and Maritime Museum, Bute Street, Cardiff CF1 6AN. Telephone: 01222 481919.

Windermere Steamboat Museum, Rayrigg Road, Windermere, Cumbria LA23 1BN. Telephone: 015394 45565.

The Staffordshire and Worcestershire Canal at Kidderminster.

10. Useful addresses

The headquarters of British Waterways is at Willow Grange, Church Road, Watford WD1 3QA (telephone: 01923 226422), where information relating to boat licences, general amenities, sailing, canoeing, angling and cruising etc can be obtained.

For licences and particulars

Rivers Great Ouse, Stour, Glen, Welland, Nene and Ancholme: Boat Registrations Department, National Rivers Authority, Anglian Region, Kingfisher House, Goldhay Way, Orton Goldhay, Peterborough, Cambridgeshire PE2 5ZR.

Avon (Warwickshire), Tewkesbury-Evesham: Lower Avon Navigation Trust, Mill Lane, Wyre Piddle, Pershore, Worcestershire WR10 2JS.

Basingstoke Canal: Basingstoke Canal Offices, Ash Lock Depot, Government Road, Aldershot, Hampshire.

Bridgewater Canal: Manchester Ship Canal Company, Estate Office, Dock Office, Trafford Road, Salford M5 2XB.

River Derwent (Yorkshire) and Market Weighton Canal: Boat Registrations Department, National Rivers Authority, Yorkshire Region, 21 Park Square South, Leeds LS1 2QG.

River Thames (between Cricklade and Teddington): Boat Registrations Department National Rivers Authority, Thames Region, Kings Meadow House, Kings Meadow Road, Reading, Berkshire RG1 8DQ.

River Wey and Godalming Navigation: The National Trust, Dapdune Lea, Wharf Road, Guildford, Surrey GU1 4RR.

The Inland Waterways Association, whose general office is at 114 Regent's Park Road, London NW1 8UQ, advocates the use, maintenance and development of the inland waterways of the British Isles and promotes the restoration to good order of every navigable waterway for use by commercial and pleasure traffic, anglers and ramblers, for water supply and land drainage, and for all aspects of multifunctional use.

11. Suggested reading

A good waterway map is essential for planning holidays, towpath walks and identifying waterway locations.

Imrays Inland Waterways of England and Wales. 10 miles to 1 inch, a detailed map of the system with enlarged plans of six areas; also shows abandoned canals.

Stanford's Inland Cruising Map. 8 miles to 1 inch; shows the waterways in relation to towns and roads.

Lockmaster Maps. 4 miles to 1 inch. Many different maps plus a Holiday Planner.

There are dozens of waterway guides, most of them inexpensive, describing both individual waterways and areas. Many are published by individual canal societies. The occasional purchase of a cruising guide will ensure that in only a short time you will have a comprehensive coverage. Initially do consider *Nicholson's O.S. Guides to the Waterways of British Waterways: South, Central* and *North.*

For specialised subjects such as fishing and wildlife there are numerous books available, but not on the subject of canal craft. To learn more about their fascinating history, development and decoration consider: *Boatyards and Boatbuilding* by Wilson, *Claytons of Oldbury* by Faulkner, *George and Mary* by Faulkner and *Life Afloat* by Wilson, all published in the Robert Wilson series; also the author's *Ethel and Angela Jane*, about the barge workings on the Calder and Hebble Navigation, and *Canal Barges and Narrowboats* and *Canal Architecture,* published by Shire Publications, and a little more expensive, *Pictorial History of Canal Craft*, published by B.T. Batsford Ltd. The domestic aspect of traditional canal life is covered in Avril Lansdell's *Canal Arts and Crafts* (Shire) and *Clothes of the Cut.*

Monthly magazines about inland waterways are: *Canal and Riverboat, Practical Boat Owner* and *Waterways World.*

For reference the following will be found invaluable: *Shell Book of Inland Waterways* by H. McKnight; *Inland Waterways of Great Britain* by L. A. Edwards; *British Canals* by Charles Hadfield.

Index

Page numbers in italic refer to illustrations.